Living Machines

The Rise of Synthetic Organisms and Bio-robots

Team Trex

Published by NobleTrex Press

For permissions and other inquiries, write to:
P.O. Box 3132, Framingham, MA 01701, USA

Contents

Introduction

Consider, for a moment, a machine the size of a grain of sand. It has no gears, no batteries, and no silicon chips. Yet, it moves with purpose. It can navigate a maze, cooperate with its neighbors to push obstacles aside, and, perhaps most startling of all, if you slice it in half, it does not break. It heals. It stitches itself back together and resumes its work within minutes.

This is not a concept from a science fiction novel, nor is it a nanobot imagined by a futurist in Silicon Valley. It is a Xenobot, a biological machine sculpted from the skin cells of an African clawed frog. It was designed by an evolutionary algorithm running on a supercomputer, but it was built by hand, cell by sticky cell. It is a robot, by any functional definition, but it is also undeniably alive.

The existence of such a creature-and we struggle to even call it a "creature"-signals that we have crossed a threshold in human history as significant as the ignition of the first steam engine or the etching of the first transistor. For millennia, humanity has categorized the world into two distinct boxes: the things that are born, and the things that are made. The born are messy, self-replicating, and evolved; the made are precise, static, and designed.

That wall has collapsed. We have entered the era of living machines.

To understand why this matters, we must stop looking at biology through the lens of the naturalist-as a collection of fascinating species to be cataloged-and start seeing it through the eyes of the engineer. Biology is the ultimate manufacturing platform. A single cell is a factory more advanced than any facility humans have ever constructed. It packs a hard drive of information in its nucleus, runs on molecular turbines that spin at thousands of revolutions per minute, and possesses a sensor suite capable of detecting single atoms. And unlike our steel and plastic factories, biology scales effortlessly. It runs on sugar and sunlight, producing complex materials with atomic precision.

For a long time, we could only admire this machinery from the outside. We were like early humans watching a bird fly, understanding that it worked but lacking the physics to build a wing. Now, however, we have cracked the chassis open. We have learned to read the source code-DNA-and, increasingly, how to write it.

This shift from reading to writing is the heartbeat of the modern biorevolution. As researchers like Craig Venter and his team demonstrated when they booted up the first synthetic cell, distinct from any natural ancestor, life is effectively programmable matter. DNA is not mystical; it is a four-letter language that instructs hardware-proteins, membranes, and enzymes-on how to behave. If you change the code, you change the machine.

But this isn't just about tweaking a gene here or there to make a tomato stay fresh longer. That is the old way. The new discipline of synthetic biology treats genetic material like lines of software code or LEGO bricks. We are moving toward a world of standardized biological parts-promoters, terminators, and coding sequences-that can be snapped together to build organisms that have never existed in nature.

The implications of this toolkit are staggering, and they touch every aspect of our existence. We are beginning to see cells not just as life forms, but as logic gates. By wiring genetic switches together, engineers are building cellular computers that can sense their environment, perform calculations, and store memories. Imagine a yogurt culture that changes color if it detects a toxin in your gut, or a plant whose leaves turn white in the presence of explosives in the soil. These are the sensors of the future: self-replicating, self-powering, and biodegradable.

As we venture deeper into this domain, we find that the "hardware" of the body is just as malleable as the software of the genes. We are no longer limited to the shapes nature provides. By combining the precision of 3D printing with the vitality of living tissue, we are learning to print flesh. We are growing mini-organs in the lab to test drugs, saving laboratory animals and humans alike from dangerous side effects. We are seeing the dawn of "smart matter"-materials that are technically alive, capable of healing cracks in concrete or adapting the weave of a fabric in response to temperature.

Crucially, this capability extends to movement and agency. The Xenobots mentioned earlier, pioneered by researchers like Michael Levin and Josh Bongard, challenge our very definitions of robotics. By stripping cells of their natural context-taking a skin cell that "thinks" it belongs on a frog and placing it in a new configuration-we discover that cells have a latent plasticity. They are willing to cooperate to build new anatomies. This suggests that the shapes of animals as we know them are just one default option on a massive menu of possible biological forms. We are learning to order off the menu.

Of course, the power to rewrite the code of life brings with it profound responsibilities and chilling risks. We cannot discuss the rise of synthetic organisms without addressing the "kill switch." If we engineer a bacteria to eat oil spills, how do we ensure it doesn't devour the plastic in our cars? If we build a virus to hunt cancer, how do we guarantee it doesn't mutate to hunt healthy cells? The engineering of life requires a safety architecture as robust as the organisms themselves-genetic firewalls, chemical dependencies that prevent escape, and a culture of transparency that brings these discussions out of the lab and into the public square.

Furthermore, the democratization of this technology changes the playing field. As the cost of synthesizing DNA plummets-following a curve that makes Moore's Law look sluggish-the ability to design life is moving from billion-dollar institutes to community labs and student competitions. The "design-build-test" cycle, once the province of aerospace and software, is now the rhythm of biology. Artificial intelligence is accelerating this loop, predicting how a strand of DNA will fold into a 3D protein structure long before it is ever synthesized. We are handing the blueprints of life to algorithms, asking them to find solutions that four billion years of evolution missed.

As we look toward the horizon, the separation between the digital and the biological blurs further. We are building "digital twins" of living systems, simulating entire organisms in silicon before committing them to carbon. We are even contemplating life that rejects the standard biochemistry of Earth entirely-organisms built with "alien" DNA bases or alternative amino acids, creating a biological firewall that prevents them from interacting with natural ecosystems.

Ultimately, this technology may be our ticket to the stars. Space travel is an exercise in resource management, and carrying enough spare parts, medicine, and food for a voyage to Mars is logistically impossible with dead matter. But if your ship is a garden-if your medicine is brewed by yeast, your structural materials are grown by fungi, and your recycling is handled by engineered bacteria-the equation changes. Synthetic biology is the enabling technology for a multi-planetary species.

We stand at a unique juncture. For four billion years, the flow of genetic information was one-way: nature wrote the script, and life acted it out. Now, we have picked up the pen. The flesh has become hardware, and the genome has become a document we can edit, cut, copy, and print.

The stories that follow are not just about science; they are about a fundamental shift in our relationship with the physical world. We are transitioning from the architects of stone and steel to the architects of living systems. The machines of the future will not just be built; they will be grown. They will not just operate; they will live.

Let's step into the lab and meet them.

Chapter 1

The Flesh is Hardware

We have long viewed life as a chaotic miracle, but under the microscope, it reveals itself as a machine built of distinct, interchangeable parts. By decoding the molecular software of DNA, we gain the unprecedented ability to debug, rewrite, and eventually compile entirely new forms of existence.

1.1 Carbon-Based Logic Gates

A smartphone feels like the most unbiological thing imaginable: a slab of glass and metal that listens, computes, and replies. Yet if you zoom far enough in, both phones and cells are busy with the same kind of work. They take in signals, they decide what those signals mean in context, and they produce actions. The difference is not that one computes and the other merely *lives*; the difference is the material. A phone computes with electrons flowing through silicon. A cell computes with molecules bumping, binding, breaking apart, and being rebuilt. In both cases, a physical system turns inputs into outputs by following rules.

Engineers use the word *gate* for the smallest unit of such rule-following. A logic gate is not a little philosopher weighing options; it is a device whose structure enforces a simple mapping. Feed it one pattern of inputs and it produces an output; feed it another pattern and it does something else. Make enough gates, wire them together, and you can build a calculator, a flight-control system, a search engine. The audacious move of synthetic biology is to say: cells already have gates, already have wiring, already have memory. The molecular details are different, but the underlying idea—that information processing is something matter can do—fits living tissue as naturally as it fits silicon.

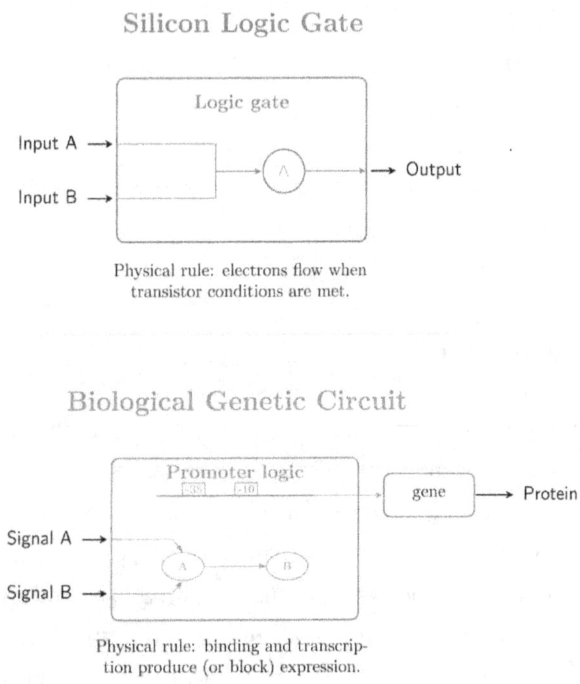

Same idea, different substrate: both map inputs to outputs via enforceable physical constraints.

To see why this analogy works, it helps to strip away the reverence that often clings to biology. The cell is not a mystical sac of life-stuff; it is an organized chemical machine with sensors and actuators. A *sensor* is any molecule that changes state when it meets a signal: a protein that binds sugar, a receptor that grabs a hormone, a DNA-binding regulator that recognizes a small metabolite. An *actuator* is any downstream effect: opening a channel, building a flagellum, secreting a toxin, turning on a stress response, dividing. Between sensor and actuator sits a logic layer, sometimes shallow and sometimes elaborate, that decides *when* an actuator should fire and *how strongly*. The logic is implemented by interactions that have the dependable bluntness of physics: molecules fit or they do not; a promoter is blocked or it is accessible; a threshold is crossed or it is not.

Promoters are a good place to start because they feel like a switch in the hand. A promoter is a DNA region that controls whether a gene is transcribed into RNA. Transcription is performed by RNA polymerase, a molecular machine that moves along DNA and builds a complementary RNA strand. Many promoters are regulated by transcription factors: proteins that bind specific DNA sequences near the promoter and either recruit polymerase or get in its way. That arrangement already looks like a gate. One input signal changes the concentration or activity of a transcription factor; the factor binds DNA; transcription goes up or down; protein output changes. If you can make the transcription factor respond to a chosen chemical, you have a programmable input.

The Central Dogma—DNA makes RNA makes protein—is often presented as a one-way street of biological destiny. In practice, it is closer to an instruction pipeline on a factory floor. DNA stores sequences; transcription makes working copies; translation turns those copies into proteins; proteins then *act back* on the world, including on DNA itself. The cell is full of loops. A protein made from a gene can turn that very gene off, creating negative feedback, a classic control motif. Two proteins can mutually repress each other, creating a bistable switch, a kind of cellular memory. A protein can activate its own gene, creating a latch that stays on even after the initiating signal fades. These patterns are familiar to anyone who has studied electronics, but they were discovered in cells first because evolution built them long before engineers named them.

A historical anecdote makes the point. In the late 1950s and early 1960s, François Jacob and Jacques Monod worked out how *E. coli* decides when to digest lactose. The lac operon is now a textbook staple, but it remains a vivid example of molecular logic. When glucose is abundant, the bacterium prefers it; when glucose is scarce and lactose is present, it turns on the genes that import and break down lactose. The decision is not made by a single molecule but by a combination: a repressor that blocks transcription unless lactose (or a derivative) is present, and an activator system that signals when glucose is low. It is not hard to translate the story into a rough Boolean form: turn on the lactose genes if (lactose is present) AND (glucose is not present). The reality is more graded than that—cells live in concentrations, not in perfect zeros and ones—but the architecture is unmistakably computational: inputs are sensed, integrated, and converted into a regulated output.

Once you start looking, you see gates everywhere. Consider bacterial chemotaxis, the way bacteria swim toward food and away from toxins. Receptors in the membrane bind attractants or repellents; a signaling cascade changes the rotation of a flagellar motor; the bacterium runs or tumbles. The details involve phosphorylation (adding phosphate groups to proteins), which acts like a reversible switch: a protein is active when phosphorylated and inactive when dephosphorylated, or vice versa. Cascades of these modifications resemble amplifiers and filters. Cells also implement something like *adaptation*: they respond strongly to a change in stimulus, then adjust their baseline so they can remain sensitive to further changes. That is control engineering, rendered in protein and water.

Synthetic biology inherits this recognition and pushes it from description to design. If a promoter and a regulator can form a gate, then gates can be assembled into circuits. The phrase *genetic circuit* is not poetry; it is a statement that DNA parts can be composed so that one gene's output becomes another gene's input. In practice the wiring is often done with transcription factors: gene A produces a repressor protein that shuts down promoter B; gene C produces an activator that turns promoter D on. You can build an inverter (NOT) by arranging for a repressor to block a promoter: more input signal produces more repressor, which produces less output protein. You can build an AND gate by making transcription require two activators, or by placing two regulatory sites such that both must be in the permissive state for polymerase to work efficiently. You can build an OR gate by allowing either of two inputs to activate expression.

One of the most striking moments in this line of work came when researchers started treating living cells as modular gate elements that could be physically arranged. In 2011, work by Jeffrey Tabor's group, including Alvin Tamsir, demonstrated genetically encoded NOR gates in *E. coli* and used chemical communication to connect colonies. The NOR gate is especially important because it is *functionally complete*: with enough NOR gates, you can build any Boolean function. In electronics, this means you can standardize on one primitive and still construct arbitrarily complex digital logic. In biology, demonstrating a robust NOR gate suggested that a similar standardization might be possible inside cells. The clever twist was to use quorum-sensing molecules—small diffusible chemicals bacteria use to talk to one another—as "wires" between colonies. Instead of soldering metal traces, you let molecules diffuse across agar. The resulting circuits were slower and noisier than silicon, but they made the analogy tangible: you could point at a Petri dish and say, without metaphorical apology, that it was implementing logic.

At this point a skeptic may object: real cells are not crisp. They do not have perfect voltage levels or clean thresholds. They leak. They vary from cell to cell. They mutate. They get tired. All true, and none fatal. Digital electronics also began in a messy world of imperfect components, and it became powerful by layering abstraction: a slightly unreliable transistor can still be used to build a reliable gate if you design carefully; gates can be composed into circuits if you define input and output conventions; circuits can be error-corrected if you invest resources. Biology's challenges are distinctive, but the strategy of abstraction still applies. When synthetic biologists speak of modularity, they are not claiming that biology is naturally modular in a frictionless way; they are trying to *make* it modular enough to engineer.

The most immediate biological friction is *context.* A promoter that behaves like a clean switch in one strain may behave differently in another because the cell's internal economy is different: polymerase levels, ribosome availability, metabolic state, DNA supercoiling, growth rate. Even the act of expressing an engineered circuit imposes a burden; it can slow growth, which changes the concentrations of everything else, which feeds back on the circuit. Noise is not just a nuisance; it can be a feature. Some cells deliberately use stochasticity to diversify responses, a bet-hedging strategy. But if you want computation to be dependable, you must design with noise in mind: add feedback, tune thresholds, avoid crosstalk, and accept that the "clock speed" of life is often minutes rather than nanoseconds.

A second friction is that biological signals are frequently continuous. The concentration of a sugar is not simply present or absent; it spans orders of magnitude. Cells respond with graded outputs: a promoter may be transcribed at 2% of its maximum, or 70%, or 99%. This is why the most honest description of cellular computation is not purely digital but *analog* or *hybrid.* The cell is a chemical analog computer that can be coaxed into digital-like behavior when thresholds are sharp and the environment is controlled. Importantly, analog computation is not inferior. It can be efficient, especially when the problem is naturally about concentrations, gradients, and probabilities. Immune systems, for example, often use analog-like integration to decide whether a signal is significant enough to trigger costly action.

There is a deeper point hiding here. When a cell "computes," it does not just represent information; it *pays* for it. To keep a protein concentration high, it must spend energy making proteins and preventing them from degrading too quickly. To keep a gradient, it must pump molecules. To maintain a memory state, it must reinforce it against the thermodynamic tendency to drift. Computation in living matter is not abstract symbol manipulation; it is a physical process embedded in metabolism. This makes cells oddly honest computers: you can often see the energy cost in their growth rate, and you can watch trade-offs play out in real time.

If genetic circuits are one way to make carbon-based gates, DNA itself offers another: computation with nucleic acids in a test tube. This can feel like a trick until you realize how much of the cell's own logic already relies on binding specificity. DNA is a polymer with predictable pairing rules: A pairs with T, C pairs with G. That predictability can be exploited not just to store information but to *process* it. In 2011, Lulu Qian and Erik Winfree demonstrated DNA strand displacement cascades that act as modular computational elements. The basic idea is simple enough to explain at a dinner party with a prop: imagine a zipper (a double-stranded DNA complex) that has a short single-stranded "tab" sticking out, called a toehold. Another strand with a matching sequence can grab that tab and begin to unzip and replace one strand with another. By designing which toeholds match which inputs, you can create gates where the presence of one strand triggers the release of another. Chains of these reactions form cascades: molecular dominoes. In such systems, “inputs” are strands you add, and “outputs” are strands that are released. No enzymes are required; the logic is implemented by sequence design and thermodynamics.

Why does this matter for thinking about cells? Because it makes computation feel less like a metaphor and more like a general property of chemistry. If you can build a logic circuit from DNA in a beaker, then a living cell—full of nucleic acids and proteins evolved to bind and switch—is not an unlikely substrate for computation; it is an obvious one. The cell’s “wires” are not copper but diffusion and binding. Its “transistors” are not semiconductor junctions but conformational changes in proteins and regulatory interactions at DNA. Its “clock” is not a quartz oscillator but the kinetics of chemical reactions and the rhythm of growth and division.

This reframing also changes how “inputs” and “outputs” are imagined. In a digital circuit, inputs are voltages on pins. In a cell, inputs can be almost anything: light (through optogenetic sensors), temperature, pH, mechanical strain, nutrients, toxins, inflammation signals, electrical potentials, even the presence of a particular RNA sequence. Outputs can likewise be diverse: fluorescence (useful in labs), secretion of a therapeutic molecule, activation of a suicide pathway, differentiation into another cell type, movement toward a signal, changes in stiffness, production of a biomaterial. The cell becomes, in principle, a programmable agent that senses the world and acts within it.

A concrete example brings the engineering instinct into focus. Suppose you want a bacterium to produce a useful compound only in a specific environment—say, the low-oxygen conditions of the gut—and only when a second signal is present, perhaps a marker of inflammation. A naive design would simply place the production genes under a promoter activated by low oxygen. But the naive design risks producing the compound in any low-oxygen setting, including places you do not want it. Logic improves specificity. You can build a circuit where the production genes turn on only when the low-oxygen sensor is active *and* the inflammation sensor is active. If either signal is missing, the output stays off. That is an AND gate implemented by biology. Replace the compound with a toxin that kills the bacterium, and you have the beginning of a containment strategy: the bacterium survives only in the intended environment. The same computational framing that helps you imagine sophisticated therapies also helps you imagine safety locks.

Of course, turning a sketch into a working living circuit is a lesson in humility. Biological parts are not mass-produced with identical parameters. Promoters have different strengths; ribosome binding sites tune protein output; proteins fold well or poorly; regulators bind with finite affinity and may cross-react. Even measuring what the circuit is doing can perturb it, since fluorescent proteins and reporters consume resources. The discipline that has grown around these problems looks increasingly like electrical engineering translated into wet lab practice: characterization of parts, standardization of interfaces, iterative debugging, and a wary respect for hidden couplings.

The phrase "carbon-based logic gate" can sound like a gimmick until you remember that silicon logic gates are also, at bottom, just matter coerced into predictability. A transistor works because the arrangement of atoms in a semiconductor makes electron flow sensitive to an electric field. A genetic gate works because the arrangement of atoms in a protein makes binding sensitive to a molecule, and because the arrangement of bases in DNA makes transcription sensitive to that binding. Both are examples of the same strategy: embed an if—then rule into physical structure.

Thinking of cells this way does not drain them of wonder; it relocates it. The marvel is not that life is beyond mechanism, but that mechanism can be so supple that it grows, heals, adapts, and reproduces. A cell is not merely a bag of reactions; it is a network of conditional operations, honed by evolution and increasingly editable by human hands. When matter can sense and decide, "hardware" starts to look less like a category and more like a spectrum, with flesh occupying a region that is slower than silicon, noisier, self-repairing, and profoundly, inconveniently alive.

1.2 Booting Up the Minimal Cell

A programmer can install a fresh operating system on a blank laptop and watch, in minutes, as the machine becomes something that can take instructions. Nothing comparable exists for an animal or a plant; life arrives preloaded, with billions of years of tinkering baked in. Yet there is a persistent temptation, both scientific and philosophical, to ask whether living things have an "OS" at all: a smallest set of instructions and machinery that, once running, can keep itself running. Not a stripped-down version of a mouse or a microbe, but a cellular core: the minimal arrangement of parts that can grow, copy itself, and keep its boundaries intact long enough to do it again.

The appeal is not only curiosity. If a cell can be understood as hardware executing genetic instructions, then finding the minimum viable hardware feels like a way to separate the essential from the ornamental. Biology, viewed from a distance, seems lavish: redundant pathways, multiple enzymes doing similar jobs, genes that appear to exist only for rare crises. Engineers eye that excess with suspicion. How much is truly required, and how much is evolutionary contingency—features that happened to be useful in a particular ecological past, but are not part of life's irreducible core?

The strongest version of the question is brutally concrete. Take a real organism, keep deleting genes, and ask after each deletion: does it still live? Do this carefully enough and you might end with a cell that resembles a freshly installed system: it boots, it runs, it replicates. It may not run well. It may crash easily. It may be fragile, slow, and strange. But it runs.

One of the most influential attempts to do exactly this came from the J. Craig Venter Institute, in a project that culminated in a cell called JCVI-syn3.0, reported by Clyde Hutchison III and colleagues in 2016. The achievement was not merely that the genome was small—although it was remarkably small, about 531,000 base pairs of DNA encoding 473 genes—but that the genome was *designed*, *synthesized*, and then installed into a living cell that could grow and divide. It was an audacious demonstration that "life's code" could be treated as an editable artifact: not just read, not just copied, but pared down and rebuilt.

Before that milestone, minimal-genome thinking had long been in the air. In the mid-1990s, when the first bacterial genomes were sequenced, researchers noticed that some microbes naturally carry very few genes. *Mycoplasma genitalium*, a tiny bacterium associated with human urogenital infections, became a celebrity of minimalism: its genome was among the smallest known for a free-living organism, and it seemed to live close to the edge, relying on a rich host environment. The logic was straightforward: if nature already produced cells with small genomes, perhaps those genomes were close to minimal, and could serve as a starting point for finding life's essentials.

But "small" is not the same as "minimal." A parasite can outsource tasks to a host and still be fully alive, but its genome is then minimal only in the context of the host's generosity. A bacterium in nutrient broth can survive with fewer metabolic genes than one living in soil. Even within a single species, a gene can be dispensable in one environment and vital in another. Minimality is always a statement with an asterisk, and the asterisk usually reads: under these laboratory conditions, with these nutrients available, at this temperature, in this chemical world.

That caveat does not weaken the project; it sharpens it. The aim is not to discover an eternal, platonic minimal genome that would function on Mars or in the ocean or in a human lung. The aim is to build a reliable baseline: a cell that contains as little as possible while still being a self-replicating system in a defined setting. Only then can one ask a deeper question: why does that baseline still contain so many mysteries?

Deletion sounds easy. In practice it is a form of biological Jenga. Each gene you remove changes the stresses on the remaining structure. Some blocks are clearly load-bearing: remove DNA polymerase and you cannot replicate DNA; remove ribosomal proteins and you cannot make proteins; remove lipid synthesis and you cannot maintain a membrane. Other blocks look optional until you realize they stabilize a process you did not even know was wobbling. A cell can sometimes limp along with missing functions, but that limping changes its internal state—metabolites accumulate, misfolded proteins pile up, membranes become leaky—and those changes can make previously "nonessential" genes suddenly essential. A deletion can also reveal hidden couplings: two genes may be individually dispensable but jointly necessary, because each backs up the other. Minimality is therefore not a simple inventory; it is a property of a network.

Hutchison's team did not reach 473 genes by rationally striking out everything that sounded unnecessary and succeeding on the first try. The story is closer to debugging than to pruning a bonsai. Early designs that seemed plausible failed to yield viable cells. That failure was informative. It meant the team's mental model of "required functions" was missing something, and the cell was an unforgiving examiner. Iteration became the method: design a reduced genome, build it, test it, learn what broke, revise the design, build again.

One of the most productive tools for learning what breaks is transposon mutagenesis, a technique that turns randomness into information. A transposon is a piece of DNA that can insert itself into a genome. If you let transposons jump around, you generate a library of cells, each with a different gene disrupted. You then ask which disruptions are tolerated. Genes that never tolerate insertion are likely essential under those conditions. Genes that tolerate insertion may be dispensable. Genes that tolerate insertion only in certain regions may contain essential domains. Importantly, this approach can also reveal "quasi-essential" genes: those that are not strictly required for a cell to survive, but are required for it to grow robustly and divide reliably. A cell may be technically alive while behaving like an unstable prototype: misshapen, slow, prone to failure. Quasi-essential genes are the difference between a machine that turns on once and a machine that can run unattended.

This distinction matters for how the minimal cell feels as a scientific object. It is tempting to imagine a minimal cell as a clean, elegant core, the biological equivalent of a well-designed microkernel. The reality is messier, and in a way more poignant. A minimal genome can still be a patchwork of necessities that do not come with tidy explanations. JCVI-syn3.0 famously retained a large fraction of genes whose functions were unknown even to experts. The cell could not do without them, but nobody could say, in mechanistic detail, what they were doing. The minimal cell therefore became an embarrassment and a triumph at the same time: triumph, because it proved that genomes can be rewritten on this scale; embarrassment, because even in a cell with fewer than five hundred genes, a great deal of the logic is still opaque.

It is worth pausing on why "unknown function" is not just a bookkeeping annoyance. In engineering, if a component is essential but its purpose is unclear, you are in a dangerous situation. You can keep the component, but you cannot confidently modify the design, and you cannot predict how the system will behave when stressed. Many genes of unknown function are not mysterious because biology is mystical; they are mysterious because their effects are distributed. They may tune membrane composition in subtle ways, chaperone other proteins into shape, maintain ion balance, or prevent toxic by-products from accumulating. Their absence may destabilize the whole system, not by removing a single obvious capability but by eroding the cell's ability to keep itself within a workable range.

That kind of function is hard to spot with traditional genetics. Remove a gene and nothing dramatic happens, until you push the cell slightly off its comfortable conditions and it collapses. Minimal-genome projects expose this harshly because they remove not only the obvious essentials but also many buffers, backups, and stabilizers. The remaining system is stripped of slack. In a slack-free system, the formerly minor roles become stark.

The minimal-cell story also complicates the notion of "booting." A computer boots because it has a separate, persistent hardware substrate and a discrete software image that can be loaded into memory. A cell does not have that separation. Its "software"—DNA sequences—is useless without a transcription and translation apparatus, but that apparatus is itself made of proteins encoded in DNA. The system is circular: genes build the machinery that reads genes. Life bootstraps itself. Every new cell begins not from a blank chassis but from a pre-existing cell that already contains ribosomes, polymerases, membranes, metabolites, and energy gradients. Even if you design a genome from scratch, you still need a living context to start it.

That is why genome transplantation, a key technique in the Venter Institute's work, was conceptually so important. The method takes advantage of the fact that DNA, once placed inside a compatible cellular environment, can take over. In essence, you provide a cytoplasmic "starter kit"—a cell body with functioning molecular machines—and you swap in a new genome. If the genome is coherent, it will begin to express itself, and over time the cell's proteins and RNAs will be replaced by those encoded by the new DNA. The identity of the cell becomes, in a practical sense, the identity of the genome. This is both an empowering idea for engineering and a reminder that "booting" in biology is always a handoff between generations of matter.

The handoff can be surprisingly rough. After the 2016 report, deeper single-cell observations made it harder to romanticize the minimal cell as neat and well-behaved. Under microfluidic microscopes, individual JCVI-syn3.0 cells did not always divide with the tidy symmetry familiar from textbook bacteria. Many showed irregular morphologies and division patterns. A population can grow even when its individuals are somewhat chaotic, as long as enough divisions succeed. But if the aim is a reliable chassis, this is unsettling. The cell is alive, but it is not stable in the way an engineer would like.

Work by Jean-François Pelletier and colleagues in 2021 explored this instability and showed that adding a modest number of genes back in—nineteen, in one variant called JCVI-syn3A—could restore more regular division behavior. The lesson was not simply "the minimal genome was missing division genes." It was subtler: cell division and morphology are emergent properties that depend on ensembles of proteins, some with unclear roles, working together. You can keep the canonical division machinery and still get bizarre shapes if supporting processes are missing. In other words, a minimal cell reveals how much of life's hardware is about *maintenance*: ensuring that core processes happen not only in principle but reliably, cycle after cycle, despite noise, fluctuations, and wear.

There is a human story embedded in these details, and it is not only about famous names. Minimal-cell work changes the daily texture of laboratory life. It forces a confrontation with biological stubbornness. In a typical genetic experiment, a failure can be blamed on a cloning error, a bad reagent, a mismeasured condition. In a minimal-genome project, failure is often real: the design is wrong because the model is incomplete. That can be emotionally taxing, because it implies that the organism is teaching you what you do not understand, and it does so by simply refusing to live. The reward, when a design finally works, is not only a publication but a rare feeling of conceptual contact: you have built a coherent self-replicating system with fewer parts than nature usually bothers to carry.

Minimal cells also invite a different kind of companion project: whole-cell computational modeling. A genome is a parts list, but a cell is a dynamic process. In 2012, Jonathan Karr and colleagues presented a whole-cell computational model for *Mycoplasma genitalium* that attempted to predict phenotype from genotype by integrating many cellular processes into a single simulation. The ambition was striking: instead of treating metabolism, replication, transcription, translation, and division as separate topics, treat them as interacting modules and ask whether a computer model can reproduce observed behavior. Even imperfectly, such models offer a new way to think about debugging life. If a minimal genome fails, is it because ATP runs short at a crucial moment? Because a metabolite becomes toxic? Because ribosomes are overcommitted? A simulation can suggest hypotheses and point to measurements. It can also reveal how much is still unknown by showing which parameters must be guessed.

The pairing of minimal genomes and whole-cell models has a certain intellectual symmetry. Minimization removes parts to see what breaks. Modeling tries to predict what breaks before the experiment. Both treat the cell as a system that can, at least partially, be understood in terms familiar from engineering: modules, interfaces, loads, failure modes. Both are humbled by the same problem: biology is dense with interactions, and many are not annotated in any database. Yet the very attempt to model or minimize forces clarity. It compels researchers to name assumptions, to quantify what can be quantified, and to accept when the organism contradicts them.

It is also impossible to avoid the ethical and practical shadow that follows minimal-cell work. Building a simpler organism can sound like making life safer, because fewer genes might mean fewer capabilities. But simplification can also make an organism more legible and therefore more engineerable. A chassis that reliably runs a designed genome is valuable for benign purposes—manufacturing drugs, sensing pollutants, producing materials—and potentially for less benign ones. Minimality does not automatically equate to harmlessness. A stripped-down cell in a rich environment can still be a powerful factory. A cell with fewer regulatory constraints may behave unpredictably under selective pressure. Even the concept of "unknown function" becomes a biosecurity issue in miniature: if an essential gene's role is unclear, it is harder to anticipate how a modified organism will behave outside the lab.

For many people, the minimal cell also touches a philosophical nerve. If you can write a genome that boots a living cell, does that mean life is just information? The honest answer is that life is not *just* information, because information does nothing without a physical substrate and a supply of energy. But information matters in a special way: a genome is a molecule that encodes constraints on future molecules. It is a chemical that builds a chemical factory that builds more of the chemical. That looping causality is not supernatural, but it is distinctive. A minimal genome makes the loop easier to see, because there is less ornament around it.

The deepest surprise of minimal-cell work may be how far "minimal" still is from "understood." Five hundred genes is not much compared to a human genome, but it is still a dense, interlocking machine whose smooth operation depends on countless molecular encounters per second. A single gene deletion can alter the shape of a cell, the timing of division, the leakiness of a membrane, the fidelity of replication. The system is not fragile because it is poorly designed; it is fragile because it is operating near the edge of what is sufficient. That edge is precisely where the outlines of the chassis become visible.

There is a final, almost domestic image that helps make sense of what it means to boot a minimal cell. Imagine inheriting a century-old house and deciding to learn how it works by removing everything that seems nonessential. You take down decorations, then walls, then insulation, then some beams that look redundant. The house still stands, improbably, but now every gust of wind whistles through, every rainstorm finds a gap, and the remaining beams creak under loads they were never meant to carry alone. You learn, in a visceral way, which parts are structural and which are comfort. You also learn that "comfort" is not frivolous: insulation and redundancy are what make a structure livable over time. A minimal cell is the creaking frame of life, still standing, still reproducing, but doing so with so little margin that the hidden work of maintenance becomes impossible to ignore.

1.3 The Cut-and-Paste Revolution

A surprising amount of modern life depends on a quiet act of copying and pasting. Not only in documents, but in medicines, agriculture, forensic testing, and the routine tracking of viruses. The act sounds abstract until you remember what is being edited: not a paragraph, but a molecule. DNA is text written in a four-letter alphabet, and for most of the twentieth century it was a text that could be read only slowly and altered only crudely. What changed was not a single miracle tool, but a sequence of inventions that turned genetics into something closer to editing. Cut here. Copy this fragment. Paste it there. Combine parts. Try again.

The revolution began, like many revolutions, with a defensive technology repurposed. Bacteria have enemies. Viruses called bacteriophages land on them like microscopic pirates, inject DNA, and try to convert the cell into a virus factory. Bacteria did not survive billions of years of that harassment by being passive. They evolved defenses that are molecular, rapid, and often elegant. One of the oldest defenses is blunt: enzymes that chop up foreign DNA. Another is subtler: a memory of past infections stored in the genome itself. That is CRISPR, and it eventually became the most versatile genetic editing tool ever devised.

But CRISPR's success is easier to appreciate if you first understand the earlier, less glamorous tools that made genetic cut-and-paste imaginable. In the early 1970s, recombinant DNA was not an industry; it was a daring idea that provoked real fear. The fear was not irrational. If you learn to move genes between organisms, you might create new pathogens or ecological disruptions. Even the scientists building the tools worried about what they were unleashing, which led to the 1975 Asilomar conference, an unusual moment when researchers publicly debated limits and containment. The technical spark behind that era was the discovery and practical use of restriction enzymes: proteins, evolved by bacteria, that recognize particular DNA sequences and cut them.

Restriction enzymes were a gift from microbial paranoia. Each enzyme recognizes a short "word" in DNA, often a palindromic sequence, and cleaves at or near it. Because the recognition is sequence-specific, restriction enzymes made DNA a substrate you could manipulate with intention. Before that, DNA was largely handled as a bulk chemical. After restriction enzymes, it could be handled as a set of fragments with defined ends. In 1978, the Nobel Prize in Chemistry recognized this development, and the award signaled something important: cutting DNA at specific sites had become a general-purpose operation, like cutting metal with a standardized tool.

Cutting, however, is only half of editing. Pasting required DNA ligase, an enzyme that seals breaks in the DNA backbone. With restriction enzymes and ligase together, researchers could do something that reads now like a children's craft project but was once astonishing: cut a gene out of one DNA molecule and paste it into another, making a recombinant plasmid. A plasmid is a small circular DNA molecule that bacteria naturally carry and copy. If you insert a gene into a plasmid and introduce the plasmid into bacteria, the bacteria can copy the plasmid and sometimes express the inserted gene. In the mid-1970s, this became the foundation of molecular cloning, and with it a new kind of laboratory routine: DNA became something you could build, propagate, and share.

The first generation of gene editing, then, was not editing a genome but editing portable DNA circles and moving them between cells. This was transformative, but it had limits that mattered. Restriction enzymes cut only where their recognition sequences exist. If you want to cut at a location that does not contain the right sequence, you either accept a messy workaround or you cannot do it. There were ways to add new sites or use other enzymes, but the core constraint remained: the editor could cut only at pre-approved addresses scattered through the DNA. The genome was not yet a document you could place a cursor into; it was a book in which you could tear pages only where perforations happened to be.

The second essential tool was not a cutter but a copier. In 1983,

Kary Mullis and colleagues introduced the polymerase chain reaction (PCR), a method for amplifying a chosen DNA segment exponentially. PCR works by cycling temperature to separate DNA strands and then allowing short "primer" sequences to bind and guide copying. The trick that made it routine was the use of a thermostable DNA polymerase, such as Taq polymerase, that can survive repeated heating. PCR turned genetics into something fast. If you wanted a particular gene, you no longer had to isolate it laboriously from large amounts of biological material. You could pull out a tiny sample, amplify the region you cared about, and have enough DNA to work with in a few hours.

PCR did more than accelerate existing workflows. It changed the emotional tone of experimentation. Copying became cheap and plentiful. That abundance enabled a culture of iteration: try a construct, test it, modify it, try again. It also enabled genetic diagnostics and forensic DNA typing, where the ability to amplify small traces of DNA is the difference between silence and a readable signal. In editing terms, PCR was the moment when you could not only cut and paste but also duplicate text at will.

Still, as anyone who has tried to assemble a complicated piece of furniture knows, a pile of parts is not yet a coherent object. Early cloning often meant one insert at a time, one restriction site at a time, with a series of intermediate plasmids. The third major step in the toolchain addressed assembly. In 2009, Daniel Gibson and colleagues demonstrated an approach now known as Gibson assembly, an elegant method for joining multiple overlapping DNA fragments in a single isothermal reaction. Instead of relying on restriction sites, you design fragments so their ends overlap by a chosen sequence. A 5' exonuclease chews back ends to create single-stranded overhangs, complementary overlaps anneal, a polymerase fills gaps, and a ligase seals nicks. It is a choreography of enzymes that produces a seamless

DNA construct with no extra scars.

Gibson assembly made it realistic to build long stretches of DNA from many pieces without depending on convenient restriction sites. It also meshed naturally with DNA synthesis, the commercial ability to order custom DNA sequences. Together, synthesis and assembly changed what "constructing a gene" meant. If you can design a sequence on a computer, order it, and assemble it quickly, the constraint shifts away from manual manipulation and toward design choices and biological function. The lab becomes a place where the bottleneck is not always technique but understanding: what sequence should you build, and what will it do in a living cell?

By the early 2010s, then, the field had a mature cut-and-paste toolkit for making DNA molecules and moving them into cells. But genomes themselves—the chromosomes that carry an organism's essential information—remained stubborn. Editing a genome usually required coaxing a cell's own repair machinery into incorporating a desired change, often at low efficiency. Researchers could insert transgenes randomly or swap genes with careful homologous recombination in some organisms, but it was rarely quick, and it often felt like negotiating with the cell rather than commanding it. What was missing was a way to make a precise break at an arbitrary location in the genome, because DNA repair is much more willing to rewrite a region when it has been cut.

This is where CRISPR enters, not as a vague promise but as a remarkably direct solution. CRISPR stands for "clustered regularly interspaced short palindromic repeats," a name that reflects how it was first noticed: odd repeating sequences in bacterial genomes separated by unique "spacer" segments. Those spacers turned out to be a memory bank of past viral infections, snippets of viral DNA stored as a warning system. When the bacterium transcribes the CRISPR region, it produces small RNAs that guide associated proteins to matching sequences in invading DNA. The proteins then cut the invader. The bacterium uses genetic memory as a targeting system.

The leap that made CRISPR an editing tool was realizing that the targeting system is programmable. In 2012, Martin Jinek and colleagues, working with Jennifer Doudna and Emmanuelle Charpentier, showed that a CRISPR-associated protein called Cas9 can be directed to cut DNA at a sequence specified by a guide RNA, provided a short adjacent motif (a PAM) is present. They also demonstrated that the natural RNA components could be simplified into a single-guide RNA (sgRNA), making the system easier to retarget. In editing terms, this was the cursor. If the guide RNA is the address, Cas9 is the blade. Change the guide sequence and the blade moves to a new location.

The elegance of the system is worth stating plainly. A restriction enzyme recognizes a fixed sequence because its protein structure is tuned to bind that particular DNA word. To change where it cuts, you would need a new protein. Cas9, in contrast, uses RNA to recognize the target by base-pairing. To change the target, you change the RNA, not the protein. This is like replacing a hard-coded search function with a user-supplied query.

The following year, 2013, several groups showed how powerful this could be in mammalian cells. Work by Feng Zhang's lab (including Le Cong and colleagues) and by George Church's group demonstrated multiplex genome engineering: using multiple guide RNAs to target several genomic sites at once. This mattered not only for speed but for the kinds of questions and applications that became feasible. Biology is full of redundancy. If you knock out one gene, another may compensate, and the phenotype may be subtle. Multiplex editing lets researchers remove several genes together, or introduce coordinated changes, revealing interactions that were previously masked.

Cas9 also arrived with knobs and variants. A double-strand break is a dramatic event, and cells respond with repair pathways that can be harnessed or that can frustrate you. One common repair process, non-homologous end joining, is quick and error-prone, often introducing small insertions or deletions that disrupt a gene. Another, homology-directed repair, can copy a provided DNA template and make a precise change, but it is less efficient and depends on cell state. By mutating Cas9 so that it nicks only one DNA strand, researchers could bias repair outcomes in some contexts and reduce certain kinds of errors. The tool was not static; it became a platform for continuous refinement.

There is a temptation, when describing CRISPR's impact, to portray it as sudden omnipotence. Real laboratories are less cinematic. Getting CRISPR to work well can still require careful guide design, delivery into cells, and rigorous checking for off-target effects. Even when the DNA cut happens exactly where intended, the cell's repair machinery is not a deterministic printer. It can produce a variety of outcomes, some desired and some not, and the distribution depends on cell type and conditions. Editing is therefore not simply *writing*; it is writing with a collaborator who has strong opinions about punctuation.

Still, the change in everyday capability is hard to exaggerate. A graduate student can now target a gene in a mammalian cell line in weeks rather than months or years, and can often do so without mastering the arcana that previously guarded genome manipulation. The democratization has consequences. It accelerates research. It lowers barriers to entry for innovation. It also amplifies the need for norms, safety practices, and thoughtful oversight, because powerful tools spread quickly and are not inherently selective about the user's intentions.

A concrete example shows how the cut-and-paste revolution interacts with the earlier idea of cells as programmable devices. Imagine an engineered immune cell designed to recognize a cancer marker and respond only under particular conditions, reducing collateral damage. That logic depends on receptors, signaling pathways, and transcriptional programs—biological circuits. Building such cells requires precise genetic edits: inserting receptor genes into safe genomic locations, knocking out genes that dampen activity, adding regulatory elements that tune expression. Restriction enzymes and classical recombination could, in principle, achieve some of this, but the process would be slow, piecemeal, and limited. CRISPR enables a more direct architecture: multiple targeted changes, in parallel, with designs that can be iterated. The cell becomes not just a natural system to be nudged but a platform that can be reconfigured.

The minimal-cell projects described earlier also rely on the broader toolchain. You cannot delete, synthesize, and transplant genomes at scale without reliable methods for assembling long DNA sequences and verifying them. Even when CRISPR is not the primary method for building a minimal genome, the culture of precise editing and rapid iteration that CRISPR embodies permeates the field. The sense that DNA can be treated as editable text encourages more ambitious designs, including designs that would have seemed reckless when every change required a bespoke protein tool.

There is a final twist to CRISPR's origin story that resonates emotionally because it is so characteristic of science: the most consequential technologies sometimes begin as obscure curiosities. The repeating sequences that became CRISPR were initially a genomic oddity, a pattern noticed by people paying attention. For a time, they were a puzzle in microbial genetics. Only gradually did the defensive logic become clear, and only then did anyone think to repurpose it. The arc runs from basic observation to mechanism to tool, and the tool then reshapes what questions can be asked.

That reshaping includes an uncomfortable category: dual-use risk, the possibility that a technique intended for benefit can be turned toward harm. Any technology that makes it easier to edit genomes makes it easier to alter pathogens, to change host range, to modulate virulence, or to evade detection. The correct response is neither panic nor complacency. It is a sober acceptance that genetic editing is now a general capability, and general capabilities require mature governance: responsible training, secure practices, clear community norms, and institutional and legal frameworks that evolve with the tools. The most dangerous posture is to treat editing as either forbidden magic or harmless tinkering. It is neither. It is craftsmanship with consequences.

Perhaps the most profound cultural change brought by the cut-and-paste revolution is that it collapses the distance between thinking and making. A sequence can be imagined, designed, ordered, assembled, inserted, tested. The cycle is not instantaneous, but it is fast enough that the feedback loop shapes intuition. Biologists increasingly think like engineers because the world now rewards that way of thinking: you can propose a design and actually build it. At the same time, biology resists being reduced to mere design. Cells respond, adapt, mutate, and surprise. Editing gives a stronger hand on the steering wheel, but it does not turn life into a simple machine.

The bacterial immune system that became CRISPR was built by evolution for a single purpose: survival against viruses. In human hands it became an editor, and that editor turned DNA from an inherited fate into a malleable medium. The most striking part is not that the medium is editable, but that the edits can be made in living systems that sense, compute, and act. In that setting, “cut” and “paste” are not just operations on molecules; they are interventions in behavior, reshaping what a cell will do when it encounters the world.

Chapter 2

Cut, Copy, Paste, Print

Biology is transforming from a descriptive science into a manufacturing discipline, treating DNA as software and cells as programmable hardware. This chapter reveals how standardized genetic parts, plummeting synthesis costs, and rigorous engineering cycles are enabling scientists to construct organisms with the precision of microchips.

2.1 Standardizing the Messy Web

A living cell can feel like a machine when you first meet it in a textbook: DNA makes RNA makes protein; proteins build structures and catalyze reactions; signals come in and responses go out. Then you open the incubator door and the neatness evaporates. Two tubes that were supposed to be identical glow at slightly different brightness. A strain that grew happily yesterday sulks today. A genetic construct that worked in one laboratory becomes temperamental when moved to another. Biology is not only complicated; it is *contextual.* Genes behave differently depending on the host cell, the nutrients, the temperature, the burden on the cell, and the rest of the DNA around them. The cell is a crowded, jostling place, more like a city at rush hour than a clean factory floor.

Engineers respond to such messiness in a familiar way: they standardize. Not by pretending the chaos is not there, but by building a set of conventions that let many people work on the same problem without each starting from scratch. The modern software world did not become productive because programmers discovered a single perfect language; it became productive because shared interfaces, libraries, and file formats made it possible to reuse work reliably. Synthetic biology has tried to do something similar with life, starting with a deceptively simple idea: if genetic sequences could be packaged as modular parts with predictable roles, then building a new biological function could become more like assembling than inventing.

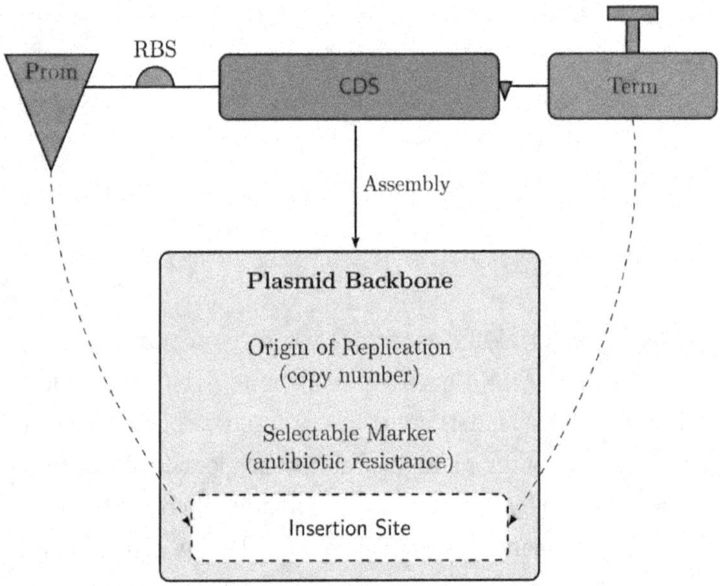

The word *part* sounds like something you can hold between your fingers, and in synthetic biology it often is. A genetic part is a defined DNA sequence that performs a recognizable function: a promoter that starts transcription (the making of RNA), a ribosome binding site that helps begin translation (the making of protein), a coding sequence that specifies a protein, a terminator that tells the cell to stop copying RNA, or a regulatory element that senses a chemical and changes gene expression in response. Once these elements are described in a common language, they become something closer to components than curiosities.

The best-known early attempt to make biology modular took shape around BioBricks. A BioBrick is not a single piece of DNA, but an approach: a part is stored and shared in a way that makes it easy to combine with other parts using a standardized assembly method. The aspiration was as social as it was technical. If many laboratories could agree on how to *package* parts, then a promoter characterized in one place might be dropped into a design elsewhere without weeks of reinvention. The Registry of Standard Biological Parts, built alongside this movement, became a communal catalogue: not just sequences, but notes about what they do, how well they work, and how they behave in specific contexts.

A historical anecdote captures the spirit. In the mid-2000s, the student-led iGEM competition became a testing ground for this modular dream. Teams arrived each summer with grand ideas—bacteria that could detect arsenic in water, cells that could form patterns, organisms that could perform logic—and a limited amount of time. The only way to build quickly was to borrow and reuse. BioBricks were meant to be the scientific equivalent of a well-stocked hardware store: you would not forge your own screws; you would pick the right size and get on with the building. The competition created a powerful incentive to treat genetic sequences as shareable parts rather than bespoke one-offs, and it pushed the community toward common standards in an unusually public way.

Yet biology resists the clean metaphor of the LEGO set. LEGO bricks are rigid and their connecting studs behave consistently across kitchens and continents. DNA parts live inside cells, and cells are not neutral containers. Even the simplest act—putting a gene on a plasmid—changes the cell's economics. Expressing a protein costs energy and raw materials; it recruits ribosomes and polymerases that might otherwise be used for growth; it can stress membranes, deplete cofactors, and trigger emergency responses. Two designs with the same DNA can behave differently simply because one strain is healthier, or because the growth medium contains a sugar that changes metabolism. Standardization is therefore always an argument with nature: a set of conventions that makes engineering *possible*, even if it never makes it effortless.

To understand what BioBricks tried to standardize, it helps to separate three layers of agreement.

First, there is physical compatibility. Early BioBrick standards defined how parts should be flanked by specific DNA sequences so they could be cut and pasted together using restriction enzymes—molecular scissors that recognize particular short sequences. The trick was to make assembly *idempotent*: when two parts were joined, the product would still look like a part, ready to be combined again. This was a clever way to avoid dead ends. It also imposed constraints: parts could not contain certain restriction sites internally, and the seams between parts left small "scars" that might or might not matter depending on what was being built.

Second, there is functional description. Knowing that two pieces of DNA can be physically joined is not the same as knowing what they will do after they are joined. A promoter is not a simple on/off switch; it has a strength that varies with conditions, and it can interact with nearby sequences. A ribosome binding site does not merely invite translation; it sets the rate at which ribosomes begin, which changes protein abundance and sometimes even the stability of the mRNA. If a part is to be reusable, it needs measurement: not only a sequence, but a characterized behavior under specified conditions.

Third, there is the representational layer: how designs are described and shared so that people and software can understand them. Engineering thrives on diagrams that compress complexity into conventions—circuit symbols, plumbing schematics, architectural plans. Synthetic biology has gradually developed such conventions (including SBOL and its visual standards) to make genetic designs legible across laboratories, companies, and software tools. Without that shared visual and digital language, even well-characterized parts remain trapped in notebooks.

The dream of standardization looks straightforward until you try to decide what, exactly, should be standardized. In electronics, voltage levels and connector shapes can be agreed upon because the underlying physics is stable. In cells, the same promoter can behave differently because the host has different transcription factors, because the plasmid copy number changes, or because the cell's growth rate shifts. These are not edge cases; they are the rule. A genetic part is less like a resistor and more like a musical phrase: it can be repeated, but its effect depends on the surrounding composition and the instrument playing it.

The response has not been to abandon standardization, but to refine it. One practical move has been to introduce reference standards. If you cannot measure promoter strength in absolute units that everyone agrees on, you can measure it relative to a reference promoter in the same experimental setup. This is a modest idea, but it matters enormously. It acknowledges that instruments, strains, and growth conditions vary, while still allowing laboratories to compare results in a way that is good enough for engineering. Measurement kits, shared protocols, and reference constructs become part of the standardization effort, not an afterthought.

Another move has been to accept that parts are not always portable as-is, and to focus on standardizing *interfaces* and *modules* rather than pretending everything is universal. In practice, many labs assemble systems from parts that have been tuned to work well in a particular organism and under particular conditions. A promoter library for *E. coli* might not behave the same way in *Bacillus* or yeast; a ribosome binding site optimized for one strain can misbehave in another. The part, in other words, carries an implicit assumption about its environment. The key engineering question becomes: can those assumptions be made explicit, and can interfaces be designed so that parts fail gracefully when moved?

This brings us to the other half of the story: the chassis.

A *chassis organism* is the host cell that carries engineered DNA. The term borrows from automobiles and aircraft: the chassis is the base platform on which you mount components. In biology, the chassis provides the replication machinery, the metabolism, the protein factories, the membranes, and the regulatory systems. A plasmid and its genetic circuit are like a device plugged into a power grid, except the grid is alive, adaptive, and sometimes resentful. The choice of chassis therefore determines how predictable the system will be.

Why has *Escherichia coli* become the prototypical chassis? Partly because it is familiar. It has been a workhorse of molecular biology for decades, with a deep library of tools: plasmids, selectable markers, promoters, strains engineered for specific tasks, and protocols honed to a kind of muscle memory. It grows quickly, tolerates being transformed (taking up new DNA), and thrives in simple media. Many of its strains are domesticated in the strict sense: they have been bred in laboratories to be easy to handle, sometimes losing traits needed for survival outside controlled environments. This domestication is one reason engineers trust it. You can make plans around its behavior.

But trust is not the same as control. *E. coli* is not a blank slate. It comes with its own regulatory networks evolved to survive unpredictable environments. When an engineered circuit is introduced, it enters a cell already full of feedback loops. The circuit may steal resources, triggering stress responses that change gene expression globally. It may produce a toxic intermediate that the cell tries to pump out. It may slow growth so much that mutant cells, which have lost or silenced the circuit, outcompete the intended design. A notorious practical problem in synthetic biology is therefore *evolutionary stability*: if a circuit imposes a cost, evolution will look for ways to remove that cost.

Standardization interacts with chassis choice in a subtle way. A part that appears reliable in one strain of *E. coli* may not behave identically in another, even if both are called *E. coli*. Small genetic differences—or differences in plasmid copy number control, or in the presence of certain repressors—can matter. The standardization project thus tends to concentrate around a few well-defined lab strains, because it is hard enough to get predictability there. This creates a tension. Standardization makes rapid progress in a narrow corridor, but biology's real applications often demand organisms outside that corridor: bacteria that live in soil, microbes that tolerate industrial conditions, cells that function in the human body, or organisms that can perform chemistry *E. coli* cannot. The question becomes whether standards can expand outward without collapsing under the weight of context dependence.

One strategy is to standardize not only the parts, but also the vector architecture that carries them. A plasmid is more than a passive container: its origin of replication determines copy number; its antibiotic resistance marker affects selection; its sequence can carry promoters or transcriptional terminators that leak into inserts; its overall size changes the burden on the cell. By making plasmid backbones modular and well-defined, engineers can swap the same genetic circuit into different backbones to tune expression levels or change host range. The plasmid becomes a platform with defined connection points, rather than a one-off craft project.

At a dinner party, the most intuitive way to picture this is to imagine writing a short piece of code. The promoter is like a function call that says "start here"; the ribosome binding site is like a setting that controls how much output you get; the coding sequence is the executable content; the terminator is the stop instruction. The plasmid backbone is the operating system and runtime: it determines whether the code can be stored, copied, and maintained. The chassis is the hardware and the living environment: it supplies power, but it also throttles, overheats, and occasionally rewrites your program.

A concrete example makes the stakes clear. Consider a simple engineered function: a bacterium that produces a visible pigment when it senses a sugar. This can be built by placing a pigment-producing enzyme under the control of a promoter regulated by a sugar-responsive transcription factor. On paper, the design is trivial: input (sugar) flips a switch (promoter), output (color) appears. In the lab, the first version may barely color at all because translation is too weak, so the ribosome binding site is swapped for a stronger one. The next version colors beautifully but grows slowly because pigment synthesis drains resources, so the promoter is weakened or the plasmid copy number reduced. Then the circuit begins to fail over days as mutants arise that shut down pigment production; selection conditions are adjusted, or the circuit is moved to the chromosome where it is harder to lose. Each step uses the same conceptual building blocks, but the parts are being tuned to the chassis and to the cell's evolutionary incentives.

Modularity helps you iterate quickly, yet it does not remove the need to negotiate with the organism.

Negotiation becomes even more important when engineered organisms might leave the lab. Standardization can help safety here too, because standardized approaches make it easier to build containment features into designs. A chassis can be chosen or engineered to be less able to survive outside controlled conditions, for example by making it dependent on a nutrient not found in nature (an auxotrophy), or by adding genetic "kill switches" that cause the cell to die if it escapes a particular chemical environment. These safeguards are not foolproof—evolution is inventive—but they can be made more robust when they are treated as reusable modules with known failure modes rather than bespoke add-ons. The same standardized mindset that enables faster building also enables more systematic thinking about risk.

There is, however, a shadow side to making biology more modular. Assembling genetic systems from standardized parts lowers the barrier to entry. That is a triumph for education and innovation, but it also creates dual-use concerns: techniques that enable benign applications can be repurposed for harmful ones. The point is not that BioBricks themselves are dangerous—most parts are mundane—but that infrastructure for easy assembly, distribution, and documentation changes what is *accessible*. Governance in synthetic biology therefore has to grapple with the fact that standardization is a form of amplification. It amplifies the capabilities of well-intentioned builders, and it can amplify others too. This is one reason why many DNA synthesis providers screen orders for sequences of concern, and why the field has invested in norms around responsible sharing and oversight.

The deepest lesson of standardizing the messy web is not that biology can be reduced to interchangeable bricks, but that a shared *grammar* makes it possible to argue productively with complexity. When a promoter is drawn with a familiar arrow symbol and measured against a reference, it becomes a communicable object: you can talk about it, compare it, improve it, and reuse it. When a plasmid backbone has a clear architecture, you can separate questions about replication and burden from questions about sensing and response. When a chassis is chosen deliberately, you can decide which uncertainties you will tolerate and which you will engineer away.

Even then, the cell keeps its personality. It surprises you, not out of malice, but because its priorities are not yours. The engineer wants a circuit to function; the cell wants to survive and reproduce. Standardization is the art of making those priorities align often enough that building becomes routine. The reward is not perfect predictability, but a new kind of leverage: the ability to take a biological function that once required artisanal craftsmanship and assemble it, piece by piece, into something that works well enough to matter.

2.2 The DNA Printing Press

A generation ago, making new DNA felt like carpentry. You started with what nature had already built, cut it with enzymes, glued it into a plasmid, then checked that the pieces had gone in the right order. The results were often ingenious, but the work was slow, tactile, and slightly anxious: a late-night gel, a shaky band, a plasmid that looked almost right. The craft had its heroes and its rites of passage, and it produced the first wave of genetically modified organisms that reshaped research and medicine. Yet it also had a ceiling. If every new genetic system required a chain of careful cut-and-paste operations, then biology would remain something you could prototype, not something you could routinely manufacture.

The shift that now defines synthetic biology is that DNA has become a medium you *write*. A design begins as a sequence on a screen, as editable as text. You can copy, search, annotate, version-control, and share it. Then, instead of hunting for a suitable template in nature, you order the sequence itself, produced by chemical synthesis. The DNA arrives in a tube, usually as a short fragment or a plasmid, already assembled into a workable form. The molecular biology that once consumed weeks can sometimes be compressed into the time it takes to decide what you want.

This change has a familiar feel because it echoes the history of publishing. Early books were laboriously copied by hand, one manuscript at a time. Printing did not make writing easy, but it altered what writing *enabled*. Once the cost and time of reproduction collapsed, authors began to think differently about what was worth composing, publishers about what was worth distributing, and readers about what was worth demanding. DNA synthesis has created a comparable discontinuity. When DNA is expensive to write, one tends to reuse, patch, and improvise. When DNA becomes cheap enough, one can plan, redesign, and iterate.

The technology underneath this "printing press" metaphor is less romantic and more precise. For decades the dominant method has been phosphoramidite chemistry: DNA is built one base at a time on a solid support, each cycle adding a nucleotide with protective groups that must be removed and reapplied in a carefully choreographed sequence. It is a triumph of chemistry and automation, and it produces the short building blocks called oligonucleotides (oligos) that laboratories rely on for PCR, sequencing primers, and assembly. Yet the method has limits: stepwise yields are not perfect, so longer sequences accumulate errors; reagents are moisture-sensitive; and long genes are usually assembled from shorter oligos rather than printed in one shot.

What has changed is not that synthesis suddenly became flawless, but that the entire pipeline around it became industrial. Instruments improved, miniaturization reduced reagent use, and companies learned how to amortize costs across enormous volumes of orders. At the same time, methods for assembling short fragments into long DNA improved dramatically. The practical consequence is that, for many tasks, the bottleneck is no longer *can we make this DNA?* but *is this the right design, and can we trust it when it arrives?*

A historical marker for this transition came from a group that treated genomes as editable files. The J. Craig Venter Institute's synthetic cell effort culminated in 2010 with JCVI-syn1.0: a bacterial genome whose sequence was stored digitally, chemically synthesized as fragments, assembled into larger pieces, and ultimately used to boot up a cell controlled by that synthetic genome. The spectacle of "creating life" drew headlines, but the deeper significance was operational. The workflow implied that a genome could be treated as a design artifact: something that could be revised in silico, manufactured in parts, and integrated into a living system. It was an extreme example, but it made a point that has since become routine at smaller scales.

The techniques that made such projects feasible also changed everyday lab work. One of the most influential was Gibson Assembly, published by Daniel Gibson and colleagues in 2009. Its appeal lies in how it makes DNA feel less like a jigsaw puzzle constrained by restriction sites and more like a word processor. If two DNA fragments share overlapping ends, a single isothermal reaction can chew back ends, allow overlaps to anneal, fill gaps, and seal nicks. The method feels almost unfairly convenient. It reduces the need to plan constructs around the presence or absence of particular enzyme recognition sites, and it makes it easy to stitch together many fragments at once.

The cultural effect of Gibson Assembly is hard to separate from the economic one. The mental model shifts. Instead of asking, "How do I find enzymes that cut in the right places?", you ask, "What sequence would I like to exist?" That sounds like a small change, but it reorganizes creativity. It encourages more ambitious refactoring: rewriting genes to remove problematic motifs, standardizing codon usage, optimizing expression, adding watermarks, and designing whole pathways as coherent systems rather than as a patchwork of borrowed parts.

Economics gives this shift its momentum. Technology observers often talk about Moore's Law, the long-running trend that computing power per dollar improves exponentially over time. Synthetic biology has its own version, sometimes called the Carlson Curve after Robert Carlson's analyses of how the cost of reading and writing DNA has fallen. The most consequential feature of these curves is not the exact slope in any particular year, but the direction of travel: DNA has moved from being a precious reagent to being a relatively inexpensive consumable. When the cost per base drops by orders of magnitude, entirely new categories of experiment become plausible. Libraries of thousands of variants, once too expensive to contemplate, become routine. Testing many designs becomes cheaper than arguing about which single design is most elegant.

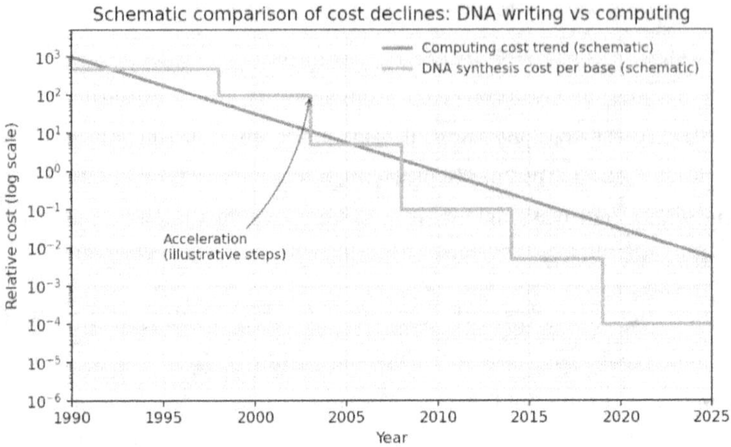

A schematic plot like this risks being too tidy, but it captures an intuition many practitioners share. DNA costs do not decline smoothly; they drop in steps as new instruments, chemistries, and business models arrive. Some costs fall faster for short oligos than for long, high-quality genes. Quality controls, turnaround times, and synthesis length limits matter. Even so, the general direction has been clear enough to reshape strategy: it is often cheaper to synthesize the thing you want than to coax it out of a pre-existing template.

The printing press metaphor also clarifies a common misunderstanding. Cheap DNA does not automatically mean easy biology. It means that one particular friction point has been reduced. In practice, cheap synthesis shifts the difficulty toward three new bottlenecks: design, verification, and context.

Design becomes a larger problem because you can now choose from an enormous space of possibilities. If a gene can be written in a thousand synonymous ways (different codons encoding the same protein), which version is best for a particular chassis organism? If a metabolic pathway can be arranged in many orders with many promoter strengths, which configuration will actually produce the molecule you want without poisoning the cell? The temptation is to treat biology like software and assume that more flexibility makes success inevitable. Instead, flexibility increases the number of ways to fail.

Verification becomes essential because synthetic DNA can contain errors, and even perfect sequences can be assembled incorrectly. The cheaper it becomes to print DNA, the easier it is to overlook the cost of confirming what you received. Sequencing—reading DNA—has become astonishingly inexpensive, and it is now routine to verify constructs before trusting them. Yet verification also has subtler meanings: confirming that a gene is expressed at the intended level, that a protein folds correctly, that the cell is not evolving away from the design. In engineered organisms, the sequence is only the beginning of the story.

Context is where the earlier theme of standardization returns in a new form. You can order a promoter, a ribosome binding site, and a coding sequence; you can assemble them flawlessly; and the resulting circuit can still behave unexpectedly because the host cell is not a blank canvas. In fact, cheap synthesis can tempt overconfidence. When it becomes easy to generate ten designs, it can feel as if one of them must work. Often that is true, but not for the reasons one might hope. It works because brute-force iteration compensates for incomplete understanding. Biology becomes tractable not because it has become simple, but because iteration has become affordable.

The analogy to computing has limits for another reason: DNA is *copied* by living systems. Once a construct is in a cell, it is replicated, sometimes at high copy number, by the cell's own machinery. A single design can therefore generate vast numbers of physical instances. This is part of biology's power: you do not merely manufacture a device; you manufacture something that manufactures itself. But self-copying introduces its own kind of cost: evolution. If a construct is burdensome, mutants that disable it will tend to take over. The printing press can produce many copies quickly, but it cannot guarantee that later copies will remain faithful without careful selection and design.

A concrete, familiar case is the synthesis and optimization of fluorescent proteins. Green fluorescent protein (GFP) began as a natural molecule from jellyfish, and it became a foundation of modern biology. Early users relied on particular sequences that worked well enough, then gradually adopted improved variants engineered for brightness, folding, and color. Today, ordering a fluorescent protein gene is trivial, but getting it to behave in a particular organism can still require redesign: codon optimization for the host, removal of unwanted motifs that trigger silencing, tuning of expression to avoid toxicity, and selection of a signal peptide if the protein needs to be secreted. The work is no longer dominated by cutting and ligating; it is dominated by choices about how a sequence will behave inside a living context.

Modern synthesis has also encouraged a new relationship to "natural" sequences. In early recombinant DNA work, a gene's sequence was often treated as a fixed fact of nature. Now sequences are frequently refactored: rewritten without changing the encoded protein, reorganized into standardized modules, or stripped of regulatory elements that make them hard to control. This can feel unsettling if one has a romantic view of biology as something best left unedited. In practice, refactoring is often a form of respect for the cell: a way of making genetic instructions legible and manageable, so that the cell is not forced to cope with strange patterns that confuse its machinery.

The industrialization of DNA writing also brings social and ethical consequences that do not fit neatly into the metaphor of printing. Printing helped spread ideas, but it also spread propaganda. DNA synthesis helps spread innovation, but it also creates dual-use risks. A world in which one can order long sequences by mail is not merely a world with faster research; it is a world that must decide how to handle access, screening, and oversight. Many synthesis providers now screen orders against databases of pathogenic sequences or toxins, and there are international efforts to standardize such practices. This is not a perfect safeguard. Screening depends on what is known, and biology has many indirect routes to harm. Still, the very fact that screening is now part of ordinary business reveals how thoroughly DNA has become an information commodity: a string of letters that must sometimes be treated like controlled information.

There is also a more personal emotional shift for researchers. The older craft of genetic engineering involved an intimate familiarity with the physicality of DNA: the smell of agarose gels, the satisfaction of a clean band, the small triumph of a successful ligation. Ordering DNA can feel like skipping the hard part, and for some it removes a sense of earned understanding. Yet it also democratizes. A small lab without a deep tradition in cloning can now participate in ambitious projects, focusing effort on design and measurement rather than on artisanal assembly. The barrier to entry drops, and the question becomes less about technical gatekeeping and more about creativity, judgement, and responsibility.

The phrase "typing code into a computer" is accurate in a narrow sense: DNA can be edited like text, and designs can be stored and shared. But it can also mislead, because biological code does not execute in a clean virtual machine. It executes in a crowded, evolving cell, whose priorities do not necessarily match the designer's. The printing press does not abolish that mismatch; it simply makes it easier to revise instructions when the first set fails.

For a long time, a great deal of molecular biology was organized around finding what already existed in nature: isolating genes, discovering regulatory sequences, hunting for enzymes in obscure microbes. That exploratory work continues, and it remains one of the pleasures of biology. What has changed is that discovery is no longer the only path to construction. A pathway can be imagined, designed from known parts, written as DNA, and tested. If it fails, it can be rewritten rather than merely tweaked. When the cost of rewriting falls far enough, the style of thinking changes: not a single heroic attempt, but a disciplined willingness to iterate.

A printing press is not valuable because it produces one perfect page. It is valuable because it makes revision cheap, and revision is how clarity is achieved. DNA synthesis has made revision cheaper in biology, and that has quietly altered what it means to engineer a living system: less an act of delicate splicing, more an act of deliberate composition, with the cell itself as the final, unforgiving editor.

2.3 Closing the Loop

Typing DNA into existence does not make living systems obedient. A cell is not impressed by the elegance of a design file, and it does not reward ambition with cooperation. It reads what it can, ignores what it cannot, and rewires what it finds inconvenient. Anyone who has built even a simple genetic circuit learns a particular kind of humility: the first version rarely works as intended, the second version works for reasons you did not predict, and the third version fails because the cell evolved a workaround while you were sleeping.

Engineering has a name for this pattern. When you cannot reliably predict how a system will behave, you stop treating each attempt as a one-off masterpiece and begin to treat it as an iteration in a cycle. Synthetic biology has adopted a version of this cycle known as Design—Build—Test—Learn, usually shortened to DBTL. The words are plain, but the concept is powerful. It turns biology from a sequence of heroic experiments into a disciplined feedback process: you design a construct, build it, test how it behaves, learn from the data, and then update the design. The goal is not to eliminate uncertainty; it is to move uncertainty into a manageable loop.

DBTL is easy to describe at a dinner party because it resembles everyday problem-solving. Imagine cooking a new dish from memory rather than a recipe. You choose ingredients (design), cook (build), taste (test), then adjust salt and heat (learn) and try again. In a kitchen, the ingredients are mostly stable. In a cell, the ingredients are alive, which means the dish may also decide to cook itself differently the next time. DBTL is the response to that stubborn liveliness.

The *Design* phase begins with a hypothesis stated in the language of parts and sequences. Perhaps a promoter is too strong and burdens the cell; perhaps a metabolic enzyme needs to be expressed earlier in a pathway; perhaps a sensor's response curve is too shallow. Modern design often happens inside software that can represent DNA constructs as editable objects, check for forbidden restriction sites, predict ribosome binding strength, and generate assembly plans. Standards such as SBOL (Synthetic Biology Open Language), first proposed in 2014 and revised into modern forms such as SBOL 3.0.0 (2020), matter here because they make designs portable between tools and between organizations. A design that cannot be shared unambiguously is like a blueprint without a scale: it may be beautiful, but it will not travel well.

The *Build* phase has been transformed by the DNA "printing press" and by assembly methods like Gibson Assembly. Yet building remains more than ordering. DNA must be inserted into cells, colonies must be screened, sequences confirmed, and constructs stabilized. Many design ideas die here not because they were logically wrong but because they are practically fragile: repetitive sequences recombine, toxic genes kill the host, or the construct is too large for the chosen plasmid. Build is where the friction between digital intent and biological reality becomes physical.

The *Test* phase is often the most revealing and the most neglected. Testing is not merely checking that a reporter glows. It is measurement: how much product is made, how quickly, under what conditions, with what burden on growth, and with what variability across cells. Synthetic biology has learned, sometimes painfully, that a single measurement at a single time point is a temptation to self-deception. Cells fluctuate. Populations contain subpopulations. A construct that looks stable for eight hours may collapse by day two. Testing therefore increasingly includes time-series measurements, single-cell assays, stress tests, and comparisons against reference standards.

The *Learn* phase is where the cycle either becomes a true engineering discipline or stays a loop of educated guesswork. Learning means taking messy data and converting it into an update: changing promoter strengths, altering gene order, rewriting codons, adding feedback control, or selecting a different chassis. It also means deciding what to trust. A measurement that cannot be reproduced is not a lesson; it is a rumor. This is where measurement standards, such as relative promoter units (RPUs) introduced by Kelly and colleagues in 2009, quietly become essential. When two laboratories can describe promoter activity in comparable units, learning can accumulate across people and time rather than evaporating when a student graduates.

DBTL became more than a slogan when laboratories began to industrialize it. The idea of a "biofoundry" emerged: a facility designed to run DBTL at scale, with robotics handling routine pipetting, automated incubators and plate readers generating standardized measurements, and software tracking every sample, protocol, and result. If a traditional biology lab is a workshop, a biofoundry is closer to a small factory—though one whose products are not finished goods but information about what works.

This is not merely about speed. Automation changes the nature of evidence. When a person pipettes by hand, subtle differences creep in: timing, mixing, temperature exposure, the very human tendency to treat one sample more carefully than another. A robot is not perfect, but it is consistent. Consistency allows finer comparisons. It becomes easier to attribute a change in output to a change in design rather than to a change in technique. In that sense, robotics is not only a labor-saving device; it is an epistemic instrument, a way of producing knowledge with tighter error bars.

A concrete example helps. Suppose a team is engineering yeast to produce a fragrance molecule, something like a monoterpene used in perfumes or flavorings. The pathway may require multiple enzymes, each of which can be expressed at different levels. Too little of the first enzyme and the pathway starves; too much of a middle enzyme and an intermediate accumulates that stresses the cell; too little of the final enzyme and the desired product never appears. The design space is immediately combinatorial. Even if each enzyme has only five plausible promoter choices, six enzymes already yield $5^6 = 15{,}625$ combinations. No single human wants to build and test that many strains by hand.

A DBTL approach reframes the problem. You choose a manageable subset to build, perhaps guided by prior knowledge. You automate the assembly of DNA constructs, transform them into yeast, and grow hundreds or thousands of strains in microplates. You test production via an assay—maybe a mass-spectrometry pipeline for accurate quantification, or a proxy reporter if direct measurement is slow. Then you learn by fitting a model: which promoter combinations correlate with higher yield, which create growth defects, which appear unstable. The next cycle designs a smarter subset rather than blindly sampling. The loop becomes a way of navigating an enormous space of possibilities without pretending you can see the whole space at once.

This is where machine learning enters with practical force. The phrase can sound grandiose, but in this context it often means something simple and useful: statistical models that can learn patterns from high-dimensional data and make predictions about untested designs. A machine-learning system does not need to understand biology in the human sense. It needs only to recognize that certain sequence features, promoter choices, ribosome binding site strengths, or codon patterns tend to produce certain outcomes in a given chassis under particular conditions. It is the kind of learning that many experienced biologists do intuitively—a feel for what will "probably work"—but translated into a form that can scale, remember, and quantify uncertainty.

A historical anecdote captures how quickly this became real rather than aspirational. In 2018, researchers reported an automated DBTL pipeline that cycled through design and testing with dramatic improvements for a pathway, showing that systematic iteration could achieve large gains over just a small number of cycles. The precise numbers are less important than the message: the loop itself can be an engine of progress. When iteration is fast and data are organized, improvement is not a lucky accident; it becomes a predictable outcome of effort.

To appreciate why prediction matters, consider the most common failure mode in synthetic biology: the design works in principle, but it collapses in practice because the cell finds it burdensome. Burden is a slippery word. It can mean metabolic cost, competition for ribosomes, membrane stress, accumulation of toxic intermediates, or activation of stress responses that ripple through the cell's regulatory networks. Burden can also be evolutionary. If a design reduces growth rate even slightly, faster-growing mutants that disable the design will eventually dominate. Many genetic constructs therefore face a double challenge: they must function biochemically, and they must be *worth keeping* from the cell's point of view.

Machine learning is increasingly used to predict burden and stability before building. For example, models can be trained on data linking sequence features to expression levels, linking expression levels to growth defects, or linking plasmid architectures to stability. In protein engineering, models can help predict whether a designed enzyme is likely to fold correctly. In circuit design, models can suggest which combinations of promoters and regulators are likely to behave robustly rather than oscillate wildly. These predictions are never perfect, but they are often good enough to focus expensive experimental work on the most promising regions of design space.

The move from hand-built loops to automated loops has also changed the geography of biology. "Cloud labs" offer a service model: designs are submitted digitally, experiments are run by robots in centralized facilities, and data are returned. In principle, a researcher could design a set of constructs on Monday, submit them, and receive results without ever touching a pipette. This decouples biological experimentation from physical location. It can also widen access, allowing groups without extensive wet-lab infrastructure to perform sophisticated experiments. Yet it creates new dependencies and new kinds of opacity. If experimentation becomes a service, the user must trust the service's protocols, calibration, and data management. The need for standards grows sharper, not weaker.

Standards sit quietly at the center of DBTL because the loop depends on continuity. A learning algorithm cannot learn from data that are incomparably measured. A biofoundry cannot automate a workflow that is reinvented each week. Even human reasoning struggles when each experiment uses a different naming scheme, a different data format, and a different set of controls. This is why design standards like SBOL and measurement approaches like reference-based units are not bureaucratic add-ons; they are the connective tissue that allows iteration to compound.

The importance of interoperability becomes obvious when biofoundries try to collaborate. A construct built in one facility should be buildable in another without translation errors. A protocol should be portable, at least at the level of its abstract steps: culture, induction, sampling, measurement. In 2019, efforts such as the Global Biofoundry Alliance helped formalize this international community, emphasizing shared practices and the idea that biological engineering could benefit from something like manufacturing standards. More recently, proposals have argued for abstraction hierarchies not just for DNA parts but for operations: breaking complex experimental processes into modular unit operations that can be composed and audited. The point is not to turn biology into bureaucracy; it is to make it possible for results to travel.

There is an emotional dimension to closing the loop that is easy to miss if one focuses only on efficiency. Many scientists are drawn to biology because of its richness and surprise. Automation can seem to drain the romance. Yet it also changes what can be discovered. When a robot can run thousands of experiments with consistent technique, it can reveal subtle structure in a system that would remain invisible in sparse data. Iteration can become playful in a new way: you can ask questions that are too large for artisanal methods. The surprise does not disappear; it is redistributed. Instead of being surprised by whether a cloning step worked, you are surprised by the shape of a response curve across thousands of designs.

Closing the loop also forces a more honest relationship with failure. In older modes of lab work, failure was often ambiguous: did the idea fail, or did the technique fail? DBTL tries to separate these, making technical execution more reliable so that conceptual failure becomes clearer. That clarity can sting. It can also be liberating. If failure is informative and iteration is fast, then abandoning a design is not defeat; it is learning in the literal sense, a reduction in uncertainty.

All of this still rests on a stubborn fact: cells are not passive platforms. Even in the most automated biofoundry, the living part of the system introduces variability. Cultures grow at slightly different rates. Gene expression pulses. Mutations arise. Contamination happens. DBTL does not abolish these, but it creates a way to cope that is both pragmatic and intellectually satisfying: treat variability as data, not as an embarrassment. Measure it, model it, design around it.

The final turn of the loop is almost philosophical. When you build, test, and learn repeatedly, you begin to see the cell less as a black box and more as a collaborator with constraints. The cell tells you, through growth rates and expression profiles and mutation patterns, what it will tolerate. A good design listens. It uses the cheapness of DNA and the speed of automation not to bully the organism into submission, but to find a configuration in which function and survival can coexist. In that coexistence, the engineered system stops feeling like a fragile trick and starts feeling like something that could live in the world: not perfectly predictable, but reliably enough to be useful.

Chapter 3

Logic Inside the Nucleus

DNA is often likened to a blueprint, but synthetic biologists view it as executable code waiting for a compiler. By stripping down natural regulatory networks and rebuilding them with standardized genetic parts, we are transforming cells into living decision engines. This transition turns the nucleus from a library of evolutionary history into a programmable logic controller.

3.1 The Switch and the Dial

A cell is awash in chemistry, yet it often behaves as if it has opinions. It commits. It remembers. It stays the course even when the world around it wobbles. A stem cell chooses to become muscle and does not drift back to being undecided; an immune cell that has been primed responds faster the next time; a bacterium that has switched into a protective state may remain stubbornly dormant. Those commitments can be described with a word borrowed from electronics: *bistability*. The idea is simple enough to explain at a dinner table. Some systems do not merely respond to an input; they settle into one of two stable states, and once settled, they tend to stay there. That staying-put is memory.

Engineers love bistability because it is the smallest unit of a digital computer: a bit. A mechanical light switch is a physical bit in your hand. Up is on, down is off. You can flip it with a fingertip, and it will hold its state all night without you continuing to press. In living cells, the finger is a molecule, the switch is DNA plus proteins, and "holding its state" means maintaining a pattern of gene expression for hours, days, or even across generations of cell division.

The striking thing is that nature already has versions of this everywhere. Developmental biology is full of switches: feedback loops in which a gene product reinforces its own production or suppresses alternatives, locking a cell into a fate. But the ambition of synthetic biology is not only to admire these switches, but to build them on purpose, with parts you can swap, tune, and reuse. The moment that ambition began to feel plausible to a broad audience came around the year 2000, when Timothy Gardner, Charles Cantor, and Jim Collins demonstrated a deliberately engineered genetic "toggle switch" in *E. coli*. The phrase stuck because it made an abstract claim concrete: cells can be wired.

To see why this matters, it helps to separate two kinds of control that are often mixed together in everyday speech. There is the *dial*, where turning a knob gives you "more" or "less": more insulin, less pigment, a stronger stress response, a weaker one. Then there is the *switch*, where the system makes a discrete choice: on or off, A or B. Biology uses both. The dial is how cells match protein levels to demand; the switch is how they avoid indecision when indecision would be dangerous. A bistable switch turns a dial-like chemistry into a yes/no commitment.

The engineered toggle switch is built from a motif so minimal it feels like a child's drawing: two genes that repress one another. Gene A makes Protein A, and Protein A binds near Gene B to prevent it being expressed. Gene B makes Protein B, and Protein B binds near Gene A to prevent it being expressed. If Protein A is high, Gene B is suppressed, so Protein B stays low, which removes repression on Gene

A, keeping Protein A high. That is self-consistent. If Protein B is high, the opposite happens, and that is also self-consistent. Between the two lies an unstable middle ground: a precarious balance where a small push tips the cell into one stable state or the other.

Right after a cell is engineered with such a circuit, it does not *automatically* end up in the same state every time. Tiny differences in molecule numbers, timing, and growth conditions can nudge one cell toward A-high and another toward B-high. That variability is not a bug; it is part of what makes bistability so powerful. A deterministic dial makes every cell in a population behave similarly; a switch can split a population into distinct subpopulations, each with its own state.

The "flip" is where the engineering becomes emotionally satisfying, because it is where intention meets mechanism. In the Collins lab's original demonstration, the circuit was designed so that chemicals called inducers could temporarily interfere with repression. Think of an inducer as a chemical glove that covers a repressor's DNA-gripping hand. Add the inducer for Protein A, and Protein A can no longer effectively repress Gene B. Gene B begins to express, Protein B accumulates, and once Protein B is high enough, it shuts off Gene A. Remove the inducer, and the new state can persist, because now Protein B is doing the holding. You have not merely stimulated a response; you have written a state into the cell.

That idea, "writing a state," is why bistable switches became a kind of mascot for cellular computing. A typical gene responds to a signal while the signal is present and relaxes when it is gone. A bistable switch, by contrast, converts a transient input into a durable record. The cell behaves less like a thermometer and more like a notepad.

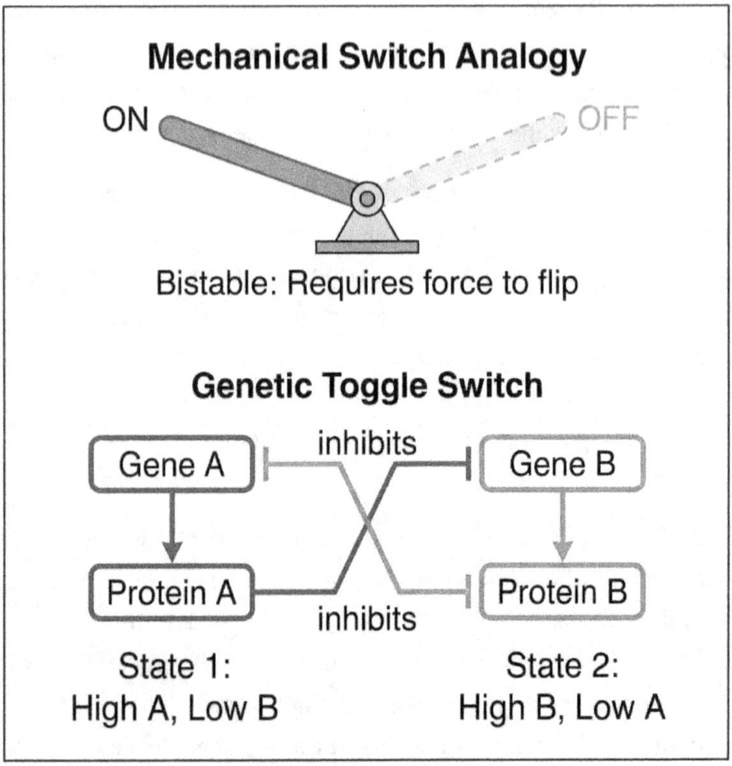

The appeal of the toggle is not only that it resembles a light switch, but that it gives a language for thinking about living machines. A circuit diagram is not a cell, but it is a promise: if you can identify parts that behave in roughly predictable ways, you can begin to compose them.

In practice, that promise meets several kinds of friction, and understanding them is part of understanding why "logic inside the nucleus" is both exhilarating and hard. The first friction is that gene regulation is not a clean on/off relay. Repression is a probability game: a repressor binds DNA, unbinds, binds again. RNA polymerase arrives, sometimes it is blocked, sometimes it sneaks through. When molecule numbers are small, chance events matter. A few extra copies of a repressor can decide whether a cell flips or stays. In electronics, a wire carries trillions of electrons and the law of large numbers makes behavior crisp. In cells, the numbers are often closer to dozens or hundreds.

The second friction is that the "parts" of biology are not manufactured to tight tolerances. Two cells carrying the same DNA can express different amounts of a protein because of differences in growth rate, ribosome availability, plasmid copy number, or local chromosomal context. Even within one cell, the effectiveness of a promoter or repressor can shift with temperature, pH, or metabolic state. That is why early synthetic biology often felt like learning an instrument: you could write down the notes, but getting them to sound right took practice.

Yet those frictions are not merely obstacles; they are also where biology offers possibilities that electronics does not. A bistable switch built from mutual repression does not require a rigid, crystalline world to function. It needs only a regime where feedback overwhelms drift. It is robust not because it is perfectly precise, but because it is self-correcting within each basin of attraction: once A is ahead, it tends to stay ahead, and small disturbances are pushed back down.

The toggle switch also opens a door to a more nuanced point: in living systems, "digital" and "analog" are rarely pure. The switch is digital in its long-term state, but analog in its dynamics. You can push gently and not flip it; push harder or longer and you might. The cell integrates the strength and duration of an inducer pulse. The bit is not a single molecule; it is a collective state of many molecules, and collective states have thresholds.

That threshold behavior becomes even clearer when you look at how biologists tune these circuits. A bistable switch is not guaranteed simply by drawing two inhibition arrows. Whether a system is genuinely bistable depends on details: how strongly each repressor binds, how quickly proteins are made and degraded, how cooperative binding is (whether repressors work better in pairs or clusters), and how "leaky" the promoters are when supposedly off. If leakiness is too high, the "off" gene still produces enough protein to prevent the other from dominating, and the system collapses into a single fuzzy state. If degradation is too slow, the switch may be sluggish or hard to flip. In other words, the difference between a decisive commitment and a muddle can be a few molecular constants.

Engineers describe this tuning with the language of *transfer functions*: how much output protein you get for a given input. A dial-like system aims for a smooth, predictable curve. A switch-like system aims for a curve with a steep slope and, crucially, *hysteresis*. Hysteresis means that the input required to flip from State 1 to State 2 is different from the input required to flip back. That asymmetry is what allows memory. It is also what makes the switch resilient to noise. If random fluctuations around the threshold could flip the state back and forth, the circuit would be useless. Hysteresis creates a gap: a safety buffer between "definitely A" and "definitely B."

A concrete way to picture hysteresis is to imagine a thermostat with a deadband. If the heater turned on at 20°C and off at 20°C exactly, a tiny draft would cause it to chatter. Real thermostats turn on at, say, 19°C and off at 21°C, creating a stable comfort zone. A genetic toggle uses feedback to create an analogous comfort zone for molecular states.

This is where the dial returns, not as the opposite of a switch but as its partner. Biologists often want a system that can behave as a dial under ordinary conditions, yet has a switch hidden inside it: a point where a gradual change produces a sudden, irreversible transition. Many cellular decisions are like that. Consider a bacterium approaching starvation. It may gradually increase stress-response proteins (dial), but at some point it may commit to a dormant spore-like state (switch). That commitment is not just a larger amount of stress response; it is a different program. The engineered toggle is a stripped-down imitation of such decisions.

Once you begin to think in these terms, you start noticing that "flipping a switch" is not always best done with a chemical. Chemical inducers are powerful, but they diffuse, they can be slow to remove, and they can have side effects on metabolism. A pulse of light, by contrast, can be delivered precisely, targeted to a region, and stopped instantly. That is why optogenetics, originally developed as a way to control neurons with light-sensitive proteins, has become an attractive control interface for synthetic gene circuits too. Light can act like a remote control for transcription factors: illuminate a cell with one color and a regulatory protein binds DNA; illuminate with another and it releases; or use light to bring two protein halves together to form an active regulator.

Seen from the outside, this looks almost magical: a petri dish of bacteria that change their gene expression pattern when you shine a patterned image onto them. Seen from the inside, it is the same old logic: an input perturbs repression, feedback amplifies the perturbation, and the system settles into a stable state that persists after the stimulus ends. Light simply replaces the chemical finger with an optical one.

There is, however, a deeper reason light-based control feels transformative. Chemicals are global. Unless you are doing microfluidics or single-cell delivery, the whole culture gets the same inducer. Light can be local. That means bistable switches can be used not only to store memory over time, but to create memory over space. Different regions of a cell lawn can be set to different states, like writing pixels.

The boundary between regions then becomes a place where biology's messiness, diffusion, and growth all conspire to blur, sharpen, or distort the pattern. The switch becomes a primitive unit of biological "display technology," and also a tool for probing how populations behave.

The historical story of the toggle switch is sometimes told as a triumph of rational design, but the more honest story includes the thrill of finding out that a living cell will tolerate your abstract idea at all. *E. coli* is not a blank circuit board. It has its own priorities: grow, divide, manage stress, keep DNA intact. When you add a synthetic circuit, you draw on the same pool of ribosomes, energy, and molecular chaperones that the bacterium needs for its own life. This is often called *metabolic burden*, a phrase that sounds clinical but has a vivid meaning: if a cell spends too much effort making your proteins, it may grow more slowly, mutate, or find ways to silence the circuit. In the laboratory, burden shows up as a quiet, frustrating drift: a culture that once behaved neatly starts to misbehave after days of growth because the "best" cells, evolutionarily speaking, are the ones that cheat.

Bistable switches are, paradoxically, both victims and beneficiaries of this evolutionary reality. They are victims because any circuit that imposes burden invites escape. They are beneficiaries because a switch can sometimes be designed so that each state is relatively self-maintaining with minimal ongoing cost. If the "memory" can be stored in a stable regulatory pattern that does not require constant high-level expression, the cell pays less rent. This insight helped motivate later approaches where memory is stored not in a continuously maintained tug-of-war between proteins, but in DNA itself through recombination: a physical rearrangement that persists even if the proteins go away. But even without recombination, the toggle switch already hints at this economy: once one repressor dominates, the other can be held off without both being produced at maximal levels.

The switch also raises a question that becomes impossible to avoid once living machines move from benchtop demonstrations to anything that touches the world: how do you make sure a cellular bit does not flip when it should not, or persist when it should not? In a laboratory flask, an unexpected state change is a nuisance. In a therapeutic microbe inside a patient, it could be dangerous. A bistable memory element can be a feature for safety (a cell permanently switches into an "off" state after completing a task), or a risk (a cell switches into an "on" state in the wrong environment and stays there).

That is why real designs rarely rely on a single switch as the sole arbiter of a high-stakes behavior. They are embedded in layers of containment: auxotrophies that prevent survival without a supplied nutrient, kill switches that trigger cell death under forbidden conditions, and regulatory designs that reduce the chance of accidental activation. Even then, the lesson of the toggle is sobering: biology can store information, and information can persist. Any organism that can remember can also misremember, and any design that can be flipped can potentially be flipped by an unanticipated input.

At the same time, the capacity for memory is exactly what makes synthetic organisms feel like a new kind of machine. A robot that forgets everything the moment you stop controlling it is not much of a robot. The genetic toggle switch is a minimal demonstration that a cell can be given a decision with consequences that outlast the trigger. That durability is what allows engineered cells to count exposures ("have I seen this toxin before?"), to record events ("did I pass through the liver?"), or to commit to a program only after a decisive signal. It is also what makes them feel, in a faint but undeniable way, like they have an inner life: a private state that you cannot see unless you ask in the right way.

A light switch clicks with a satisfying finality because it moves through a point of instability and lands in a stable groove. The engineered genetic toggle does the same, only its groove is carved from feedback, its click is molecular, and its consequence is a cell that behaves tomorrow as if it still remembers what happened today. That is a small thing to build, two repressors glaring at each other across a loop of DNA, yet it is enough to turn chemistry into choice.

3.2 Boolean Logic in Wetware

A single genetic switch is already a small marvel: a cell that can be nudged into one of two stable moods and then hold that mood as it grows and divides. But a bit by itself is just a memory cell. Computation begins when bits *interact*: when a cell waits for two things to be true before it acts, or when it responds to either of two dangers, or when one signal cancels another. That vocabulary—AND, OR, NOT—comes from George Boole's nineteenth-century attempt to turn reasoning into algebra. The surprise is not that biologists borrowed the words; it is that living cells can be made to obey them.

To say a cell implements an AND gate means something precise. There are two inputs, each of which can be present (1) or absent (0), and there is an output that should be on only when both inputs are 1. If the inputs are "sugar" and "toxin," the output might be a fluorescent protein, a drug, or a self-destruct program. The cell becomes a tiny bureaucrat: it refuses to stamp the form unless both boxes are ticked.

This feels intuitive when the inputs are simple chemicals, because gene regulation already works by "permission slips." Many promoters—stretches of DNA that control whether a gene is transcribed into RNA—need more than one condition to be met. One protein might recruit RNA polymerase, another might open the local DNA structure, a third might remove a repressor. Nature uses such combinatorial control to ensure that genes turn on only in the right tissue, at the right time, under the right conditions. Synthetic biology repurposes the same logic, but with a different goal: not to mirror nature's intricacy, but to build circuits that can be reasoned about.

The physical meaning of logic in cells is easiest to see by imagining a promoter as a docking station with multiple slots. Input 1 produces a transcription factor that can bind to slot A; input 2 produces another that can bind to slot B. A gene fires only when both slots are occupied, because only then is the transcriptional machinery assembled strongly enough. That is an AND gate built from molecular crowding and cooperation. OR gates can be built by giving the promoter multiple alternative ways to activate: slot A or slot B will do. NOT gates are even closer to everyday biology: a repressor protein binds DNA and blocks expression unless something removes it.

A dinner-party version of this might sound like a set of bouncers at a club. The AND club checks two IDs. The OR club lets you in if you have either a ticket or your name on the list. The NOT club lets you in unless you are wearing the wrong wristband. What makes the cellular version different is that the bouncers are made by the club itself, and the rules are implemented by binding events that happen at random in a warm, crowded, vibrating molecular soup.

This randomness is why building larger logic circuits inside a single cell quickly becomes less like wiring a breadboard and more like organizing a city. Each additional "gate" typically means additional proteins that must be produced, folded, and kept from interfering with one another. Each additional DNA element must be placed somewhere in the genome or on a plasmid, where neighboring sequences and local chromatin structure (in eukaryotes) can change its behavior. The components are not passive. A transistor in a laptop does not demand amino acids, but a synthetic repressor does. A wire does not hog ribosomes, but a messenger RNA does.

The term *metabolic burden* is sometimes presented as an engineering nuisance, yet it is also a moral fact about living machines: computation costs life. A cell has a limited budget of energy and molecular attention. If a circuit consumes too much of that budget, the cell grows slowly, becomes stressed, accumulates mutations, or simply evolves to silence the circuit. This is one reason why cells are not likely to become general-purpose computers in the silicon sense. They are better thought of as decision-makers embedded in metabolism: specialized logic, paid for by biochemistry.

Early demonstrations of genetic logic often used transcriptional cascades: one regulator controls another, which controls another, which finally controls an output. Cascades are conceptually straightforward—stacking NOT gates is a classic way to build arbitrary logic—but they are slow (each layer requires transcription and translation) and noisy (variability multiplies down the chain). As circuit depth increases, timing becomes uncertain and the output becomes less crisp. A cell also has no clean equivalent of an insulated copper trace. Regulators can bind off-target sequences; promoters can respond to the cell's own proteins; small molecules can have side effects. Logic, in wetware, is always negotiating with context.

One historical pivot point came when researchers began to focus on a single logical operation that is, in a strict mathematical sense, enough to build all others: NOR. A NOR gate outputs 1 only when both inputs are 0. If either input is 1, the output is 0. This seems awkward until you remember that biology is naturally good at repression. If an input can produce a repressor, then "input present" can mean "output blocked." NOR takes that inversion seriously and turns it into a design strategy: build one reliable repressing gate, then wire many of them together.

In 2011, Alvin Tamsir, Jeffrey Tabor, and colleagues demonstrated a genetically encoded NOR gate in *E. coli* and used it as a building block for more complex logic. The cleverness was not only in making a gate, but in realizing that wiring inside a single cell was not the only option. They used quorum-sensing systems—chemical languages bacteria use to talk to one another—as "wires" between spatially separated colonies. One patch of bacteria produced a signaling molecule; another patch sensed it and changed gene expression accordingly. By arranging colonies on a plate in different patterns, they could implement different logical functions. The petri dish became a circuit board, and the wires were made of diffusing molecules.

That demonstration had a theatrical quality that matters in science. The "computer" was visible: you could look at a plate and see logic happen as patterns of fluorescence. It also carried a practical lesson. If you spread computation across many cells, each cell can carry a smaller burden. You can reuse the same cell type in multiple locations. You can reduce unwanted interference by keeping incompatible parts in different strains. Most importantly, you can exploit a fact that electronics tries to avoid but biology cannot: space. Diffusion, gradients, and distance are not annoyances; they are computational resources.

Even if one keeps the computation inside a single cell, the idea of a functionally complete set of parts pushes synthetic biology toward something that looks, uncomfortably, like software engineering. A logic circuit is a specification: "express gene X if and only if conditions A and B are present and condition C is absent." The messy work is translating that specification into DNA sequences that behave that way in a particular strain grown in a particular medium at a particular temperature. As designs grew more ambitious, the field began to treat genetic circuits less like handcrafted sculptures and more like compiled programs.

This is where tools such as Cello matter, not because they magically make biology easy, but because they make the difficulty legible. In the mid-2010s, work led by researchers including Alec Nielsen and Bryan Der showed that a high-level logic description could be compiled into a DNA design by choosing from a library of characterized genetic gates and simulating how they would behave together. The analogy to electronic design automation is not a marketing gimmick. It acknowledges a hard-earned truth: circuits fail not only because the idea is wrong, but because the parts interact in unexpected ways. Characterization—measuring how a promoter responds to a regulator, how steep the response curve is, how leaky the "off" state remains—becomes the equivalent of knowing the electrical properties of a component. Compilation becomes the process of selecting parts whose behaviors match and whose crosstalk is minimal.

The word *insulation* appears often in this context, and it carries an almost psychological meaning. A well-insulated gate behaves similarly regardless of what is upstream or downstream, regardless of where it sits in the genome, regardless of what other proteins are competing for the cell's resources. Perfect insulation is impossible in a living cell, but better insulation is the difference between a circuit that can be reused and a circuit that works only once in one lab's hands. Some strategies for insulation are physical (using genetic "spacer" sequences, terminators, and ribozyme elements to reduce interference), some are regulatory (using orthogonal transcription factors that do not interact with the host), and some are architectural (distributing gates across cells or using DNA recombination rather than continuous transcriptional regulation).

Recombinase-based logic deserves special attention because it feels like a conceptual leap. Transcription-factor logic is like using people to hold doors open: the output stays on only while the right proteins are present at the right concentrations. Recombinase logic is like remodeling the building. Recombinases are enzymes that cut and rejoin DNA at specific sites, flipping or excising segments. If a recombinase flips a DNA segment containing a promoter, the gene downstream can be turned on permanently; if it excises a terminator, a gene can be unleashed; if it inverts a coding sequence, it can change which protein is made. The key is that the state is stored in DNA architecture itself. Even if the recombinase goes away, the rearranged DNA remains.

In 2013, Piro Siuti, John Yazbek, and colleagues demonstrated how recombinases could be used to implement logic and memory in living cells in a highly systematic way, including the full set of two-input Boolean functions. The practical implication is that a cell can be "programmed" not only to compute an output from inputs, but to *record* that computation in a durable form. The philosophical implication is more unsettling: information can be written into the genome as a direct consequence of environmental signals. The cell becomes not merely a sensor but an archivist.

Recombinase logic also reshapes the burden question. A deep transcriptional cascade requires continuous production of regulators to hold a state. A recombinase circuit might require a short pulse of enzyme expression to flip DNA and then very little ongoing cost. That can make circuits more stable over long growth periods, because evolution has less incentive to silence a system that is not constantly draining resources. It can also make circuits more dangerous if misused, because irreversible DNA changes are, by design, hard to undo. The same feature that makes memory reliable makes mistakes persistent.

So far, all of this sounds like digital logic trying to colonize biology. But cells are not naturally digital. Many of the most useful biological computations are analog: proportional responses, graded secretion, smooth adaptation. A bacterium does not benefit from an all-or-nothing appetite for sugar; it benefits from a metabolism that ramps up as sugar becomes available. An immune system does not benefit from producing zero cytokines until a threshold is reached and then producing a thousand; it benefits from nuanced control that avoids runaway inflammation. The dial is not inferior to the switch; it is simply different.

Some researchers have therefore explored synthetic circuits that compute analog functions more directly, using feedback to shape input-output curves. In 2013, work by Ramiz Daniel, Jacob Rubens, and collaborators argued that insisting on crisp digital gates can be inefficient in resource-limited cells, and demonstrated circuits that perform analog computations such as ratiometric sensing and arithmetic-like operations using relatively small numbers of components. The point is not that cells will replace calculators, but that biological "computation" often means transforming signals in ways that are useful for survival or therapy, not producing binary answers.

This tension—digital clarity versus analog economy—shows up in practical design choices. If a therapeutic microbe must decide whether it is in the gut or in the bloodstream, a switch-like response is attractive: only one environment should trigger drug release. But if the goal is to adjust dosage based on inflammation level, a dial may be safer. Overly digital behavior can create cliffs: just below threshold, no drug; just above, too much. A dial can provide smoother control, though it may be harder to guarantee that "off" is truly off.

Real circuits often blend the two. A system might use analog sensing to integrate noisy inputs and then pass the result through a thresholding module that acts like a switch. Or it might use a switch to commit to a mode ("therapy enabled"), and within that mode use analog control to tune output levels. These hybrid designs echo a familiar pattern in everyday technology: a thermostat is a switch driven by an analog sensor; a camera's autofocus is a digital decision built from analog signals.

If all this sounds abstract, it is worth grounding it in a concrete, almost homespun example. Imagine engineering a bacterium that lives in soil near plant roots and releases a protective compound only when two conditions are met: the plant is present (detected by a root exudate) and a pathogen is present (detected by a toxin or stress signal). An AND gate prevents wasteful production when only the plant is present; an OR gate might allow response to either of two pathogen markers; a NOT gate could prevent release when oxygen levels suggest the bacterium has drifted away from the root zone. None of these gates requires the bacterium to "understand" plants or pathogens. It only needs molecular detectors coupled to gene regulation. The logic is a bridge between chemistry and intention.

This bridge has social consequences. The more predictable genetic logic becomes, the more it invites a certain kind of scaling: libraries of reusable parts, standardized measurement, design tools that lower barriers. That democratization can be creative and beneficial, but it also carries a dual-use shadow. A circuit that can recognize a combination of signals and trigger an output is, in the wrong hands, a recipe for harm. The same tools that make it easier to design a microbe that releases a drug only in a tumor microenvironment could make it easier to design a microbe that activates in response to other cues. This is not a reason to abandon the work, but it is a reason to treat "programmable cells" as a technology that deserves governance, not just enthusiasm.

On a practical level, biosecurity and biosafety shape circuit design in ways that are rarely visible in glossy demonstrations. Researchers build in dependencies—nutrients not found outside the lab, or genetic "kill switches" that trigger self-destruction under certain conditions. They choose chassis organisms with long histories of safe use. They avoid designs that could plausibly increase environmental fitness. And they test, repeatedly, for failure modes: mutations that disable the safety mechanism, unexpected activation of outputs, leakage that undermines a gate's truth table.

The need for truth tables in biology is itself revealing. When an engineer says a circuit is an AND gate, the statement is an aspiration: the output should match the four rows of the table. In a living cell, each row is a distribution, not a single value. Even when average behavior looks correct, some cells may miscompute. There is a deep lesson here about what it means to compute in a material that reproduces, mutates, and varies. Reliability often comes not from making every cell perfect, but from making populations robust: redundancy, thresholds, and collective averaging.

Boolean logic in wetware is therefore less like a miniature laptop and more like a council that reaches a decision by rules. Individuals can be mistaken, but the group can still act sensibly if the rules are chosen well. A promoter that requires two transcription factors is a rule. A recombinase that permanently flips a DNA segment is a rule. A compiler that chooses parts based on measured behaviors is a rule. Together they allow the designer to say, with increasing confidence, "this cell will act only when these conditions are satisfied."

There is something quietly profound about that phrase "only when." It captures the ethical appeal of cellular logic as much as the technical one. The dream is not merely to make cells do things, but to make them *selective*: to release a potent molecule only in a specific microenvironment, to kill a cell only when it has crossed a boundary, to record an exposure only when a pattern of signals makes it meaningful. Selectivity is what turns power into usefulness. It is also what turns the cell from a passive vessel into a conditional agent, one whose behavior depends on a small internal calculus written in DNA and protein.

When a colony of engineered bacteria glows only at the intersection of two diffusing chemicals, it can look like a parlor trick. The deeper point is that a cell has been given a rule of inference, and the rule is enforced not by silicon and voltage but by binding affinities, transcription rates, and the finite patience of metabolism. The logic is real because the consequences are real: a gene is expressed or not, a toxin is produced or not, a memory is written or not, and the cell lives with the result.

3.3 The Internet of Living Things

A lonely engineered cell is an impressive gadget, but it is also a fragile one. It has only so many molecules it can spare for computation, only so many promoters and repressors you can cram into its genome before the parts begin to interfere, and only so much patience before evolution edits your design into something more convenient for survival. The easiest way around these limits is disarmingly old-fashioned: share the work. In electronics, distributing computation means networks, packets, and protocols. In biology, it often means a chemical whisper that becomes a chorus.

Bacteria have been doing this long before humans learned to solder. Many species release small molecules into their surroundings and sense how much of that molecule has accumulated. When there are few cells, the signal disperses and its concentration stays low. When cells become dense, the signal builds up until it crosses a threshold, triggering a coordinated change in gene expression across the population. This is quorum sensing, a phrase that makes microbial life sound like a town meeting. The metaphor is apt. A single bacterium has no way to measure "how many of us are here?" directly. But it can measure "how much of our shared signal is in the water?" and treat that as a proxy.

What happens at that threshold often looks like a switch at the level of the whole community. *Vibrio fischeri*, a marine bacterium famous from biology textbooks, waits until it is concentrated inside a squid's light organ before it turns on its luminescence genes. Light is expensive; making it in open seawater would be wasteful. In the shelter of the host, high density is guaranteed, and coordinated light becomes useful. This story is sometimes told as a charming naturalist anecdote, but it contains a design principle: a population can behave like a bistable element, flipping from "quiet" to "loud" as density crosses a boundary. The bit is no longer inside one cell. The bit is the crowd.

Synthetic biologists recognized early that quorum sensing could be used as something more than a biological curiosity. It could be a wiring method. If one engineered cell could send a molecule and another could receive it, then the molecule could stand in for a wire carrying a signal between circuit modules. The dream of a "cellular computer" inside one chassis organism could be replaced by something more scalable and, in many cases, more reliable: a multicellular circuit in which different strains play different roles.

This shift changes what "logic" means in practice. Inside a single cell, the chief difficulty is interference: proteins that bind the wrong promoter, resource competition that changes expression levels, noise that causes occasional misfires. In a distributed circuit, some of those difficulties become easier. You can keep incompatible genetic parts in different strains. You can tune each module separately. You can insulate modules not only with genetic tricks but with literal physical distance. Yet new difficulties appear. Signals diffuse. Cells move or grow into one another. The timing of messages depends on growth rate and geometry. A wetware network lives in space, and space has its own habits.

A vivid demonstration of this "living network" style came from work by Tamsir, Tabor, and colleagues in 2011. They built logic gates in *E.*

coli and connected them using orthogonal quorum-sensing systems, treating chemical signals as interconnects. Colonies on an agar plate were arranged like components on a circuit board. By changing the arrangement, they could implement different logical functions, including ones that are awkward to build inside a single cell. The plate itself became part of the computation. Distance acted like a resistor or a delay line. Diffusion replaced copper.

It is hard to overstate how conceptually liberating this is. A person trained in electronics often thinks of wiring as a necessary evil: something that should be as short, lossless, and interference-free as possible. In biology, the "wire" is a molecule that spreads, degrades, and is eaten by the environment. Rather than fight this, multicellular designs use it. A diffusing signal creates a gradient, and a gradient can be interpreted. The receiver can be programmed to respond only within a certain concentration window, so that it turns on not everywhere the signal exists, but only where the signal is "just right."

This concentration-window behavior lies behind one of the most beautiful ideas in the field: engineered pattern formation. In 2005, Subhayu Basu, Yoram Gerchman, and collaborators built a sender—receiver system in bacteria using a quorum-sensing signal (AHL, an acyl-homoserine lactone). Sender cells produced the signal; receiver cells carried a genetic circuit designed to respond in a band-pass way, meaning it activated only at intermediate concentrations. On a plate, that translates into a ring. Near the sender, the concentration is too high and the receiver stays off. Far away, the concentration is too low and the receiver stays off. In the middle lies a halo where the concentration falls within the activation window, and the receiver turns on, often marked by fluorescence. A microscopic rule becomes a macroscopic pattern, not by any cell knowing the whole picture, but by each cell applying the same local logic to its local chemistry.

At a human level, ring patterns have an emotional pull. They look intentional, almost decorative, like biological crop circles. But their real importance is that they show computation distributed across a field. The "function" is not located in a single genome or a single intracellular circuit; it is located in the interaction between signal diffusion and thresholding. When you begin to see it that way, an engineered bacterial lawn starts to resemble a kind of chemical screen: an array of identical processors whose outputs form an image.

This style of design also addresses a persistent problem from intracellular logic: scale. In a single cell, each additional gate adds burden and potential crosstalk. In a multicellular system, you can build complexity by arranging modules rather than adding parts. A colony that acts as a NOR gate can be reused multiple times, just as an electronic designer reuses a standard chip. A signal can be broadcast to many receivers, like a wireless transmission. The price is less precision and more dependence on environment, but the payoff is a circuit that can be expanded without asking any one cell to carry the whole plan.

Recent work has pushed this idea further, showing that spatial arrangement can itself be a programmable parameter. Rather than building ever more elaborate genetic wiring, one can rely on diffusion fields and geometry to implement logic. If two sources of signals sit at different positions on a plate, the intersection regions where their concentrations exceed thresholds can form shapes that correspond to logical combinations. Move the colonies, and the logical function changes, even if each colony's internal genetics stays the same. Computation becomes architectural.

This is a subtle but radical shift in how one thinks about "design." If a circuit is a sequence of DNA letters, then design is molecular. If a circuit includes spatial layout, then design becomes something like urban planning. The city blocks are colonies. The roads are diffusion paths. The traffic rules are quorum-sensing thresholds. A designer can trade genetic complexity for spatial complexity, which is sometimes easier to manage experimentally and sometimes closer to how biology actually works in tissues and biofilms.

These distributed systems also illuminate why synthetic biology has increasingly embraced population-level approaches. A single engineered cell is unreliable partly because it is an individual. Individuals vary. Populations average. When a function depends on the collective behavior of thousands or millions of cells, random fluctuations in any one cell matter less. Quorum sensing itself is a population-level noise filter: it prevents a response until there is enough collective evidence to justify the expense. In that sense, quorum sensing is not merely a communication channel; it is a consensus protocol.

Once the language of networks appears, other metaphors follow. Researchers have begun to talk about "molecular communication" more broadly: cells sending information using molecules, other cells decoding those messages, and hybrid systems in which electronic devices interface with biological networks. The Internet of Bio-Nano Things (IoBNT) is one such umbrella idea, imagining heterogeneous networks that include engineered microbes, implanted sensors, drug-delivery particles, and external controllers. The phrase can sound like science fiction, but the underlying ingredients are familiar: sensing, signaling, computation, actuation. What changes is scale and context. A molecular packet is not a string of bits traveling at near light speed; it is a cloud of molecules diffusing, degrading, binding, and sometimes being intercepted by unintended receivers.

This brings the discussion back to an uncomfortable but necessary point: connectivity amplifies both capability and risk. A distributed system can coordinate drug release across a population of therapeutic microbes, making treatment more effective and reducing side effects. The same coordination could, in principle, help an organism persist or spread in ways that complicate containment. A population that can sense its own density and flip into a protected state might be harder to eliminate. A strain that can communicate with other strains could exchange signals in unforeseen ways. The "internet" metaphor is not only about sophistication; it is also about attack surfaces and unintended interactions.

Safety-minded designers therefore take communication seriously as a controllable resource, not a free bonus. One can choose signaling systems that are orthogonal to those used by natural microbes in the target environment, reducing the chance that wild organisms will eavesdrop or interfere. One can build dependencies so that signals degrade unless a particular synthetic nutrient is present. One can include kill switches that respond to the absence of a laboratory-supplied molecule. One can design systems where communication is one-way, limiting feedback loops that could run away. None of these is perfect, but they reflect a mature understanding: once cells can coordinate, they can also surprise.

Coordination also changes what "intelligence" looks like. A single bacterium does not have a nervous system, yet a bacterial community can display behaviors that feel eerily collective: forming patterns, dividing labor, mounting synchronized responses to stress. Engineers can harness these behaviors to build swarms that do work: detect toxins over a wide area, produce a compound in a timed pulse, or build materials with structured properties. The intelligence is not a ghost in the machine. It is an emergent property of simple rules repeated across many agents.

A concrete example makes this less mystical. Imagine engineering a population where only a small fraction of cells act as "scouts," sensitive to a toxin, while the rest act as "responders," capable of neutralizing it but only when alerted. Scouts detect toxin and release a quorum-sensing signal. Responders detect the signal and turn on a detoxifying enzyme. If the toxin is absent, responders stay quiet and do not waste energy. If the toxin appears in a localized region, only scouts in that region signal, and only nearby responders activate strongly, because diffusion creates a local cloud. The population behaves like a spatially aware sensor—actuator system without any single cell knowing where the toxin is. The map is written into concentration gradients.

One can take the same idea into medicine, with all the extra ethical weight that implies. A proposed therapeutic microbe might need to detect a combination of gut-specific cues, inflammation markers, and oxygen levels, then coordinate a response so that drug release is not a trickle but a well-timed burst. If a microbe produces a molecule that dampens inflammation, too little may do nothing and too much may suppress immune defenses. A synchronized pulse can be more controllable. Quorum sensing offers a way to do synchronization without external hardware, using only local communication. The difference between a swarm and a crowd, in this context, is the difference between noise and purpose.

It is tempting to imagine these systems as miniature versions of human technologies, as if bacteria are tiny smartphones exchanging messages. The analogy helps up to a point, and then it misleads. Unlike digital networks, biological networks are inseparable from metabolism. A bacterium cannot "turn off its radio" without changing its protein production. Unlike electronic devices, cells reproduce, and reproduction changes network topology. Unlike routers, cells die, mutate, and evolve. A biological internet is therefore always a moving target: a network whose nodes are alive.

That liveliness is not only a complication; it is also the reason the idea is worth pursuing. A living network can repair itself. It can expand into new space. It can adapt to slow changes in environment. A swarm of engineered cells can, in principle, persist where a single engineered cell would fail, simply by being many and by being connected. It can also fail in new ways, because self-repair and self-expansion are close cousins of persistence.

In the end, the most convincing image of a living internet is not a diagram of nodes and edges, but a dish on a lab bench: a field of bacteria, each one following simple genetic rules, collectively producing a pattern that no single cell contains. The computation is not hidden in a black box. It is diffusing in plain sight, written in gradients, thresholds, and the suddenness with which a quiet population can, at a shared signal level, decide to light up together.

Chapter 4

Growing the Machine

When biology scales up from the single cell to tissues and materials, it stops being solely a genetic puzzle and becomes an architectural one. We venture here into the domain of the tangible, where 3D printers layer living cells into functional shapes and concrete heals its own cracks. As the physical complexity of these creations rises, the barrier to entry plummets, transforming synthetic biology from an elite discipline into a garage-friendly pursuit driven by plummeting DNA costs.

4.1 The Scaffold and the Printer

A single human cell is perfectly content on a flat surface. Give it a warm incubator, a bath of nutrients, and something to cling to, and it will spread itself out like a starfish on glass. Under a microscope, this can look like success: the cell divides, the dish fills, the experiment produces a clean readout. For much of modern biology, that flat world has been a feature rather than a bug. A Petri dish makes life legible.

But bodies are not legible in that way. Bodies are thick. They are layered, folded, threaded through with tubes, and built from materials that push back. The moment bioengineers try to grow something that has the *shape* and *function* of living tissue—a patch of cartilage that bears weight, a slab of muscle that contracts, a piece of liver that detoxifies—the old virtues of the dish become liabilities. Cells that behave beautifully as a two-dimensional carpet can behave strangely, or simply die, when asked to live in three dimensions.

Two problems appear almost immediately. The first is mechanical: soft living matter slumps. The second is metabolic: thick living matter starves. Together they explain why "growing the machine" turns out to be as much about scaffolding and plumbing as it is about genes.

Consider the mechanical problem first. Many of the materials that cells like to live in are, from an engineer's point of view, barely materials at all. Hydrogels—water-swollen networks of polymer chains—are popular because they resemble the body's own extracellular matrix, the fibrous, gelatinous environment that cells inhabit in real tissues. A hydrogel can be made from collagen, gelatin, alginate, hyaluronic acid, or synthetic polymers tuned to feel "soft" in the way that tissue feels soft. Cells sense that softness. They pull on their surroundings and decide, in part, what kind of cell to become. Too stiff, and they act as if they are on bone; too slack, and they may fail to organize.

The trouble is that what is good for a cell is often terrible for a free-standing object. A cylinder printed from a very soft gel behaves less like a plastic part and more like toothpaste: it holds its shape for a moment, then relaxes into a puddle. Print a bridge or an overhang and gravity wins. Even printing a simple vertical wall can be hard, because each new layer squashes the one below. This is not a minor inconvenience. If you cannot hold shape, you cannot build thickness. If you cannot build thickness, you cannot build organs.

Early tissue engineering tried to solve this by borrowing from craft: pour a gel into a mold, seed it with cells, and hope the cells remodel the gel into something robust. It worked for some thin tissues and simple geometries. Skin equivalents are an instructive example. A sheet of cells is, in some sense, the natural unit of skin. A thin construct has a short diffusion distance for oxygen and nutrients, and it can be nourished from the surface like a lawn. This is why skin was one of the first clinically useful engineered tissues. It is not because skin is easy, but because its architecture cooperates.

Most tissues are not so cooperative. Cartilage, for instance, is thick and load-bearing. Heart muscle is thick and rhythmic, and its cells must align and communicate electrically. Liver is thick and chemically busy, with different zones performing different metabolic tasks. The more a tissue resembles a small organ, the more its cells demand an environment with internal structure. They need cues about where to attach, when to stop migrating, and how to arrange themselves. They also need a way for nutrients to reach their interior and for waste to leave.

This leads to a second and deeper constraint: diffusion. Oxygen and many nutrients can move through water-rich gels, but they do so slowly. In real tissues, cells are rarely very far from a capillary. The vasculature is not just a delivery service; it is the condition of possibility for thickness. Without blood vessels, animal tissue becomes a thin film.

If a cluster of cells grows in the middle of a gel with no internal circulation, the cells on the outside consume the oxygen and glucose first. The center becomes hypoxic, then acidic, then toxic. Cells there may switch to emergency metabolism, stop functioning normally, or die, leaving a necrotic core. A spheroid of tumor cells in a lab dish often develops this pattern on its own, which is one reason spheroids are used as crude models of tumors. Yet for an engineered organ, a dead core is not a feature. It is a failure mode.

So the problem of three-dimensional tissue is really the problem of coupling *form* and *flow*. Cells must be placed in space, but that space must be threaded with supply lines.

Bioengineers discovered, sometimes painfully, that a printer is not a magic wand. A 3D printer can place material in three dimensions, yes, but the material must survive the act of printing and the hours and days after. When the "ink" is living cells suspended in a gel—a *bioink*—printing becomes a negotiation among competing demands.

The printer wants an ink that behaves like a solid the moment it leaves the nozzle. The cells want an ink that behaves like a gentle bath. The printer wants viscosity and yield stress so the filament does not spread. The cells want low shear stress so their membranes are not torn and their internal machinery is not scrambled. The printer wants rapid crosslinking—the chemical or physical process that turns a liquid into a stable gel—so that layers stack. The cells want crosslinking that is non-toxic, uniform, and mild in temperature, light exposure, and reactive chemistry.

One way to appreciate this tension is to picture what happens inside the nozzle. The bioink is forced through a narrow opening. The narrower the nozzle and the faster the flow, the higher the shear forces. Shear is an invisible violence: it does not look dramatic, but it can rupture cells or trigger stress responses. Cells can survive some shear; blood cells endure it routinely. But many cell types used in tissue engineering are fragile, and the printing process can select for the hardiest survivors in ways that subtly change the biology.

The engineering vocabulary here can feel cold—rheology, viscosity, yield stress, crosslinking kinetics—yet the underlying question is intimate. How much abuse can a living system take while still becoming itself?

The field has responded with a series of clever compromises. One family of compromises changes the ink. Some bioinks are designed to be *shear-thinning*: they flow under pressure but quickly thicken when the pressure is released. Others are granular or microgel-based, behaving like a pile of tiny soft beads that can be extruded and then annealed into a more continuous gel. The idea is to decouple printability from final stability: let the material be easy to push through the nozzle but stable once deposited.

Another family of compromises changes the environment. Instead of printing into air—where gravity immediately pulls on every strand—you print into a supportive medium. This approach has a satisfying physical intuition. If you have ever piped icing onto a cake, you know that the air offers no help; the icing must support itself. But if you squeeze a soft material into a bowl of thick pudding, the pudding holds it up. For a bioengineer, the pudding is a support bath.

A now-classic example is the method called FRESH, introduced in the mid-2010s by a group led by Adam Feinberg and colleagues, building on ideas in embedded printing. The acronym stands for "Freeform Reversible Embedding of Suspended Hydrogels," which sounds like a spell until you translate it into ordinary language: print soft gel into a temporary gel that behaves like a solid when left alone, but flows when the nozzle moves through it. The support bath cradles the printed structure as it forms, preventing collapse. After printing, the bath can be melted or washed away, leaving the intended shape behind.

The elegance of this idea is that it turns the mechanical problem sideways. Instead of demanding that the bioink be strong enough to stand up on its own, you let it be biologically friendly and outsource the strength to a removable scaffold. That opens a much wider range of printable materials, including those that more closely mimic the body's own soft matrices.

There is a practical, almost folk-engineering charm to the story as well. The early demonstrations were notable not only for what they printed but for how accessible the setup could be. A technique that works only on bespoke million-dollar machines stays in a few laboratories. A technique that can be adapted to relatively simple extrusion printers and open-source control software has a different social trajectory. It can spread.

Yet even with support baths, printed tissue faces the second problem: feeding the interior. The moment you can print a thick lump of living hydrogel, you have also printed a thick lump with an oxygen problem.

In the body, vasculature is hierarchical. Large arteries branch into smaller arterioles, then capillaries, then venules, then veins. The geometry is not decorative; it is a solution to constraints. Large vessels minimize resistance over long distances. Capillaries maximize surface area for exchange. If you want to keep a cubic centimeter of tissue alive, you need something like that branching network, or at least a substitute that lets fluid pass close enough to every cell.

This is where the language of "plumbing" becomes more than a metaphor. Microfluidics—the engineering of fluid flow through tiny channels—provides a toolbox for building such networks. The channels can be molded into gels, printed as sacrificial structures and later removed, or formed by printing around them. They can be connected to pumps that perfuse culture media through the tissue, delivering oxygen and nutrients and carrying away waste.

Perfusion does something else, too: it introduces force. Flowing fluid exerts shear stress on the cells that line channels. In blood vessels, endothelial cells expect this; they use it as information. Shear stress helps them align, tighten their junctions, and maintain a barrier. So a perfused channel can encourage not only survival but differentiation into vessel-like structures. A static gel is silent; a flowing channel speaks.

A concrete example makes this vivid. Imagine a bioprinted "liver chip" designed for drug testing. Hepatocytes, the liver's metabolic workhorses, are arranged in a 3D matrix. Nearby, microchannels carry culture medium. The goal is not to create an organ you could transplant, but a miniature tissue that behaves enough like liver to predict whether a drug will be toxic. Without channels, the hepatocytes in the interior would downshift or die, and the tissue would behave more like a dying liver than a healthy one. With perfusion, the tissue can sustain function longer and show gradients of oxygen and nutrients that resemble real liver zones—a critical feature, because the liver's metabolism is spatially organized.

That word *gradient* matters. In real organs, not every cell sees the same environment. Cells nearer to an artery see more oxygen than cells nearer to a vein. Nutrient concentrations change with distance. Hormones arrive in pulses. Waste products accumulate locally. These spatial differences create physiological specialization. Trying to grow tissue that behaves like an organ means embracing controlled inequality. The Petri dish, by design, erases gradients. An organ depends on them.

This is one reason the older dream of simply "adding growth factors" to a uniform gel often disappoints. Growth factors are signaling proteins that tell cells what to do. If you pour them in like dye, you may get a uniform but unnatural instruction. What cells often need is not just the presence of a signal but its *placement* and its *timing*. The engineering challenge becomes choreographic: where should the signal originate, how quickly should it diffuse or be degraded, and what should the cell experience as it moves?

Scaffolds help with this choreography. A scaffold is not merely a physical support; it is an information landscape. Its stiffness can vary from region to region. Its chemistry can present different binding sites. It can release signals slowly, or only when cells apply force, or only when a certain enzyme is present. In the body, the extracellular matrix is not a passive sponge; it stores growth factors, exposes them when remodeled, and guides cell migration through both mechanical and chemical cues. Designing a scaffold means deciding which parts of that sophistication you can afford to mimic, and which parts you must leave to the cells themselves.

The first generation of tissue engineering scaffolds were often made from biodegradable polymers familiar from sutures and implants. They provided shape and strength, then degraded as cells laid down their own matrix. This concept remains powerful, but printing pushes it further. A printed scaffold can be anatomically specific: shaped to fit a defect, porous in a deliberate pattern, stiff where load is needed and open where vessels should enter. The printer can create internal architecture that would be impossible to carve by hand.

But printing also introduces new failure modes. If crosslinking is done with light—photocrosslinking—the dose and wavelength can harm cells, damage DNA, or generate reactive chemicals. Light may not penetrate evenly through thick constructs, leading to uneven curing: a stable shell with a soft interior, or a pattern of hard and soft regions that was not intended. Even when the chemistry is biocompatible, the physics can betray you. A channel designed on a computer may partially collapse during printing, or become clogged by swelling, or trap bubbles that later block flow. Small defects matter because they propagate: a tiny blockage changes shear stress downstream, cells respond, the lining becomes irregular, the channel narrows further.

These details can sound fussy until you remember what is at stake. In a plastic part, a small void may be cosmetic. In living tissue, a small void can be a dead zone.

The most illuminating historical anecdote here predates modern bioprinting but captures the underlying predicament. In the early days of organ transplantation, surgeons learned that an organ is not merely a mass of cells; it is a vascular object. You could not simply move tissue from one place to another and hope it "took." You had to reconnect the blood supply. The invention and refinement of vascular anastomosis—surgically joining blood vessels—was as crucial to transplantation as immunosuppressive drugs. A kidney without blood flow is not a kidney; it is a wet sponge of cells in crisis.

Bioprinting confronts the same lesson, but in reverse. Instead of reconnecting existing vessels, it tries to create a structure that is ready for perfusion from the beginning. Some strategies aim to print channels large enough to be perfused immediately, then encourage smaller vessels to grow inward over time. Others attempt to print sacrificial inks that later dissolve, leaving behind fine channels. Still others print multiple cell types, including endothelial cells, so that a vessel lining forms in place. None is a universal solution, because each tissue has different demands for scale, mechanical strength, and biological maturity.

A subtle point often missed in popular accounts is that vascularization is not simply a matter of drawing a branching tree and printing it. Real vasculature develops through a mix of genetic programs and physical constraints. Flow shapes vessels, and vessels shape flow, and the network remodels in response. Designing a vascular tree is therefore not only a geometry problem but a dynamical one. The resistance in each branch, the pressure needed to drive flow, the tendency for certain regions to become stagnant, the way cells sense and respond to shear—all of these determine whether the printed network remains open and functional.

This is why the metaphor of "printing an organ" is both inspiring and misleading. Printing can place material, but it cannot by itself guarantee *development*. Living tissues mature. Cells reorganize. The matrix is remodeled. Patterns sharpen. In many cases, the printed object is a starting scaffold for a biological process that must be coaxed, not commanded.

The most persuasive demonstrations of progress tend to be those that respect this partnership between fabrication and growth. A printed cartilage construct, for example, may be designed with pores that encourage nutrient penetration and allow cells to deposit their own matrix. Over time, the construct becomes less like a printed gel and more like cartilage because the cells do the slow work of collagen organization. The printer provides the initial architecture; biology provides the long-term reinforcement.

This partnership also clarifies why simple size is a poor measure of success. A thick printed block is not automatically closer to an organ than a thin one. What matters is whether the interior is alive, whether the cells are in the right places, whether they receive the right mechanical and chemical cues, and whether the structure can be perfused in a stable way. A few millimeters of well-fed, well-organized tissue can be more technologically meaningful than a centimeter of mostly dead gel.

There is, finally, a philosophical shift hidden in all this engineering. The Petri dish encouraged a kind of biological idealism: if you understood the right molecules, you could control the cell. Three-dimensional tissue forces a more embodied view. Cells are not only chemical factories; they are mechanical and spatial creatures. They listen with their membranes. They measure stiffness with their cytoskeletons. They read gradients as carefully as they read genes. To build living machines at useful scales, you must design environments that speak the cell's native language: texture, tension, proximity, flow.

A scaffold, in that sense, is not merely a support. It is a temporary architecture of persuasion. A printer is not merely a dispenser. It is a way of placing that persuasion into space, with enough precision that millions of cells can agree on what they are collectively becoming.

When those two roles align—when the printed structure holds long enough for cells to weave their own strength, and when the channels carry enough nourishment for interior life to remain competent rather than desperate—a remarkable inversion occurs. The question stops being whether soft matter can be printed like plastic. The question becomes whether plastic-like discipline can be relaxed enough to accommodate growth. The most successful constructs are not those that resist change, but those that anticipate it: scaffolds that expect to be remodeled, channels that expect to be lined, and printed patterns that leave room for the tissue to finish the job in its own time.

4.2 Matter That Breathes

A living thing is easy to recognize when it is shaped like one: a dog, a tree, a person lying very still and then suddenly not. It becomes harder when life is smeared thinly across something that is otherwise familiar and inert. A damp stone can host a film of algae. A wooden fence can carry a small ecosystem of fungi. A kitchen sponge, left too long, becomes a miniature swamp. None of these objects is *alive* in the way an animal is alive, but none is purely dead matter either. They occupy a borderland where the word "material" starts to feel inadequate.

Engineers, for a long time, have preferred their materials to be indifferent. Steel does not have moods. Concrete does not learn from experience. Plastic does not decide, mid-use, to turn into something else. This predictability is the moral foundation of structural engineering: you want beams that behave the same way tomorrow as they did today, and you want the failure modes to be dull and calculable.

Synthetic biology complicates that comfort. Once cells can be treated as programmable systems—not perfectly programmable, but programmable enough—it becomes tempting to stop thinking of them only as things you grow *inside* devices. Why not make the cells part of the device? Why not build materials that are not merely compatible with life, but constituted by it?

This is the core idea behind what many researchers now call *Engineered Living Materials*: hybrids in which living cells provide some active function, while an accompanying matrix—a gel, a polymer, a fibrous scaffold, even sand—provides the shape and the mechanical body. It is a deceptively simple definition, and it hides a profound shift. The emphasis moves from building organs that imitate biology to building matter that borrows biology's tricks: sensing, responding, healing, growing, and adapting.

The phrase "engineered living material" can sound like marketing until you hold the alternatives in mind. Ordinary engineered materials are usually designed for a static job: bear a load, conduct electricity, block water. If they crack, you replace them. If they wear, you lubricate them. If they foul, you clean them. Living systems, by contrast, are designed by evolution to persist in a world that is constantly trying to damage them. Skin closes wounds. Bone remodels under stress. The immune system patrols and repairs. Even microorganisms form biofilms—sticky communal layers—that protect them from heat, desiccation, and antibiotics. Life's distinguishing habit is not just metabolism; it is maintenance.

Materials that can maintain themselves feel like science fiction because they would rewrite the economics of infrastructure and the habits of everyday repair. The first time a bridge "heals" after a freeze-thaw cycle without a crew and a jackhammer, the story ceases to be about clever microbiology and becomes a story about politics, insurance, and trust.

Self-healing concrete is an instructive place to begin, partly because it sounds outlandish and partly because it answers a very ordinary problem. Concrete cracks. It cracks because it shrinks as it cures, because it expands and contracts with temperature, because the steel reinforcement inside it corrodes and swells, because the ground moves, because trucks are heavy, because time is undefeated. Many cracks are hairline, and a hairline crack is not dramatic. Yet hairline cracks become highways for water and salts. Once water reaches the steel rebar inside reinforced concrete, corrosion begins, rust expands, and expansion makes more cracking. A small defect becomes a structural narrative.

The microbial idea is to recruit a biological reaction that produces mineral—essentially, to let bacteria act as microscopic masons. Many bacteria can cause calcium carbonate (calcite) to precipitate under the right chemical conditions. If those bacteria can be kept alive (or at least viable) inside the concrete, and if they can be activated when a crack forms and water enters, they can deposit calcite in the crack, sealing it. The crack becomes a wound, and the material grows a scab.

The details matter. Concrete is an alkaline, hostile environment, and the mixing and curing process can kill cells. This is why many designs rely on spore-forming bacteria: microbes that can enter a dormant state, enduring dryness and harsh chemistry, then waking when water returns. The bacteria also need a food source. In some experiments, nutrients are encapsulated alongside the spores in small pellets or microcapsules that protect them during mixing. When a crack opens, it breaks capsules or allows water to seep in, dissolving the protective shell and feeding the microbes. In the following days or weeks, mineral precipitation fills part of the gap.

Even described plainly, the approach has an almost mythic appeal: stone that heals itself when wounded by rain. It also forces a practical question that engineers cannot avoid. How long should the bacteria survive? If the material heals only once, it may still be useful. If it heals repeatedly, it is more valuable. But if living microbes remain active indefinitely in a structural material, what does that mean for long-term stability, for byproducts, for the rest of the ecosystem?

Here the romance meets the ledger. Some microbial pathways for calcite precipitation, such as those involving urea breakdown, can produce ammonia as a byproduct, raising environmental and indoor air concerns depending on the application. Different bacterial strains, different encapsulation chemistries, and different concrete formulations change the balance among healing efficiency, byproduct formation, and longevity. A successful laboratory demonstration is not the same as a material you can deploy on a bridge in Minnesota and forget for thirty winters.

Yet the conceptual leap remains important even where the implementation is imperfect. Self-healing concrete is not only a gimmick. It is a declaration that a building material can be treated as a habitat, and that the habitat's residents can be given a job.

The same declaration can be made with gentler materials. A fabric that responds to sweat, for example, might contain engineered microbes that detect salt concentration or pH and release a pigment or a fragrance. A packaging film might contain living cells that sense oxygen ingress and change color when food is no longer sealed. A wall coating might include organisms that capture carbon dioxide and deposit minerals, slowly hardening over time.

These examples are easy to imagine and hard to build. When you embed life into matter, you inherit all the unruly needs of life. Cells require water. They require energy. They require a tolerable range of temperature. They do not appreciate ultraviolet light. They mutate. They die. They can also, if poorly constrained, proliferate in places you did not intend.

So engineered living materials become a discipline of controlled domestication. The matrix must protect the cells from drying out, from mechanical stress, from predators, and from competition with wild microbes. At the same time, it must allow the cells to sense the environment and exchange molecules with it. It must not poison them, and it must not be poisoned by their metabolites. It must keep them where you want them, which is a containment problem as much as a materials problem.

Hydrogels return here in a different guise. In bioprinting, a hydrogel scaffold helps animal cells hold shape and survive. In engineered living materials, hydrogels often serve as microbe-friendly matrices: they retain water, allow diffusion, and can be tuned chemically to keep cells immobilized. A hydrogel can be cast as a coating, printed as a structure, or mixed with fillers like sand to create a composite. The resulting object is neither a pure gel nor a conventional solid; it is a hybrid with a living phase and a nonliving phase.

There is a reason the word *composite* appears so often in this literature. A composite material gains properties from combining constituents: fiberglass combines glass fibers and resin; reinforced concrete combines cement and steel. Engineered living materials extend this logic by adding a constituent that is not a passive ingredient but an active agent. The living component can respond to stimuli and produce outputs that a dead material cannot: enzymes, pigments, adhesives, fibers, minerals, antimicrobial compounds.

The most persuasive demonstrations of this approach often use microbes rather than animal cells. Microbes are hardy, diverse, and metabolically inventive. They can live in extremes, switch into dormancy, form spores, and survive without the precise conditions that mammalian cells demand. They also have shorter generation times, which makes them easier to engineer and to scale. If engineered living materials are to leave the laboratory, microbes are likely to be the first emissaries.

One striking branch of work aims to create *living building materials*: blocks or panels that incorporate photosynthetic microorganisms such as cyanobacteria. Cyanobacteria are ancient architects of the planet's atmosphere; they helped oxygenate Earth long before animals existed. Some can induce mineral precipitation, binding particles together. In living building material experiments, researchers have mixed sand with a hydrogel and inoculated the mixture with cyanobacteria. The cells, fueled by light and carbon dioxide, contribute to cementation by encouraging mineral formation and by producing extracellular substances that help bind the matrix.

The appeal is not only that the material can be grown with low-temperature processes—a sustainability advantage over conventional cement production, which is energy-intensive and carbon-emitting—but that the material can, in some versions, be regenerated. If a block can serve as a "seed" for more blocks, the manufacturing model shifts. Instead of shipping tons of clinker and aggregate, you could imagine shipping a starter culture and growing materials locally from widely available ingredients.

Even if this remains a niche or speculative pathway, it illuminates what synthetic biology adds to materials science. Traditional materials engineering treats manufacturing as a one-way path: raw ingredients become finished parts. Biology offers cyclicity: growth, repair, reproduction. Those words are unsettling when applied to building blocks because they imply agency. Yet the agency is, in practice, bounded. A living building material does not decide to walk away. Its agency is more like yeast in bread: it is active within a constrained environment, performing a biochemical role that changes the material's structure.

A useful historical analogy is the long-standing craft of fermentation. Humans have used living microbes to make stable materials for millennia: bread, beer, cheese, soy sauce. Fermentation is, in a sense, an engineered living process that produces nonliving products. The microbes do the work, then the final material is often baked, boiled, aged, or otherwise stabilized. The novelty of engineered living materials is not that life can make useful stuff; it is that life can remain embedded and functional *after* manufacturing, continuing to sense and respond.

That continuation is where both the promise and the unease lie.

To understand the promise, consider what responsiveness could mean in practice. A coating on a pipeline could sense a small leak by detecting hydrocarbons or a change in pH, then respond by producing a polymer that plugs the leak, or by generating a visible signal. A filter could sense the presence of a toxin and upregulate enzymes that degrade it. A wound dressing could sense bacterial infection and locally release antimicrobial peptides. These are not miracles; they are extensions of what microbes already do in nature, made more specific by genetic circuits that link sensing to action.

The unease is equally grounded. Living components can evolve away from their engineered function if that function is costly to them. A microbe engineered to produce a helpful polymer might, over generations, accumulate mutations that reduce production and increase growth rate, outcompeting its more dutiful siblings. In a controlled bioreactor, engineers can manage this with selection strategies and regular re-inoculation. In a wall panel on a building, evolution becomes a maintenance problem.

There is also the question of containment. In the lab, microbes are kept in flasks and incubators. In the world, they encounter rain, dust, insects, curious children, and the microbial ecosystems already living everywhere. Engineered living materials therefore force designers to think like ecologists and safety engineers at the same time. How do you prevent a strain from escaping? How do you prevent it from exchanging genes with wild microbes? How do you make sure it dies when it should, survives when it must, and stays dormant when it is not needed?

Some of the most practical containment strategies are almost disappointingly mundane: encapsulation, physical barriers, limiting nutrient availability, and designing the matrix so that cells cannot easily leave. There are also genetic strategies, such as engineering microbes to require a nutrient not found in nature (auxotrophy), so that they cannot thrive outside their intended environment. These safety locks are never absolute—biology has a habit of finding loopholes—but layered strategies can reduce risk.

Reliability is the other hard wall. Engineers want guarantees. Biology offers probabilities. A bacterial population embedded in a material may behave differently depending on humidity, temperature, and nutrient diffusion. A product that works beautifully in a controlled test chamber might behave erratically on a construction site. In engineered living materials, the environment is not a nuisance variable; it is part of the system.

This is why the most honest vision of living materials is not that they will replace conventional ones everywhere, but that they will occupy specific niches where responsiveness is worth the complexity. Medical devices, environmental remediation, and certain high-value infrastructure components may justify the added uncertainty. A self-healing crack in a remote installation is not merely convenient; it might prevent a disaster. A sensor that is also a repair mechanism might reduce inspection burdens. A coating that changes color to show stress before failure could save lives.

There is, finally, a more intimate reason living materials capture the imagination. They blur a boundary that many people carry unconsciously: the boundary between the world we inhabit and the world that inhabits *us*. Buildings are usually treated as dead shells that keep nature outside. Yet buildings already host life: mold in damp corners, bacteria on door handles, microbial clouds in ventilation systems. Engineered living materials make this visible and deliberate. They ask whether we want to treat the living presence in our constructed environments as contamination to be sterilized away, or as a resource to be managed, shaped, and perhaps trusted.

Trust is not a technical variable, but it will decide much of what happens next. A hospital may not want wall coatings that contain live microbes, however cleverly constrained. A homeowner may be uneasy about a self-healing countertop that "wakes up" when scratched. Regulators will ask what happens in a fire, in a flood, in a landfill. Insurers will ask what can be certified. The history of technology suggests that what is technically possible is only the beginning; what is socially acceptable is a second invention, often harder than the first.

Still, the central idea remains stubbornly compelling. Materials have always been a way of freezing human intention into matter: a bridge is a decision made durable; a bottle is a choice about how to hold liquid. Living materials add a new ingredient to intention: the capacity for ongoing response. They do not merely *resist* the world; they participate in it.

A crack that seals itself is not just a repaired defect; it is evidence that the boundary between fabrication and maintenance has become porous. A wall panel that slowly mineralizes under light is not just a greener brick; it is a different relationship between time and structure.

The material is no longer a finished statement. It is a sustained act, kept going by cells doing what cells do best: turning energy into organized change.

4.3 The Garage Biologist

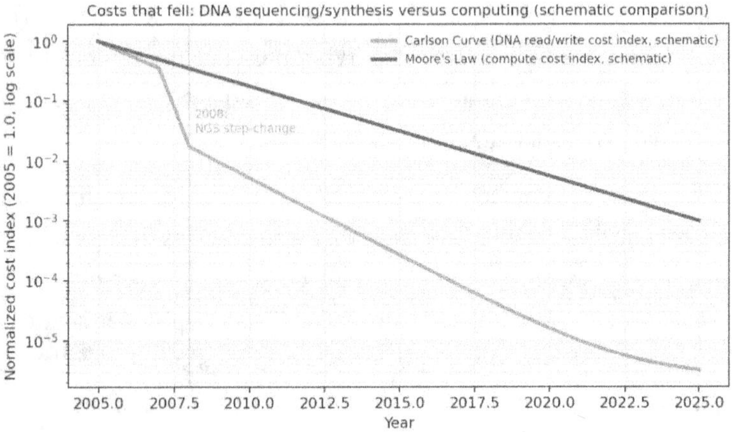

There is a particular kind of confidence that comes from being able to do something in your own space, on your own time. It is the confidence of the teenager with a soldering iron, the hobbyist with a lathe, the musician with a laptop and a cheap microphone. Synthetic biology, which once seemed inseparable from large institutions and large budgets, has been sliding toward that kind of confidence for more than two decades. Not because cells have become simpler, but because the tools for reading and writing DNA have become cheaper, faster, and more routine.

The cultural shorthand for this trend is the image of the garage: improvised benches, second-hand equipment, open-source software, experiments done after work. Like any shorthand, it is partly myth.

Biology still needs sterile technique, proper waste disposal, temperature control, and training. A garage is not automatically a laboratory. Yet the metaphor persists because it captures a real shift in who can participate. When the cost of experimentation drops, the gatekeeping role of capital weakens, and other forces rush in to take its place: curiosity, community, ambition, and sometimes recklessness.

The engine behind this shift is economic, and it can be drawn as a curve. In computing, people talk about Moore's Law: the historical pattern in which the number of transistors on a chip doubled roughly every couple of years, bringing cheaper and more powerful computation. Synthetic biology has its own version, often called the Carlson Curve after Rob Carlson, who argued that the cost of sequencing and synthesizing DNA has been falling even faster than computing costs. The details depend on which technology you track and which year you start counting, but the lived experience in laboratories is clear enough: tasks that once required major grants and months of waiting now fit on a credit card and a calendar.

The collapse of DNA sequencing costs is the most dramatic illustration. For much of the twentieth century, reading DNA was slow and artisanal, done in specialized facilities. Then, around the mid-2000s, high-throughput *next-generation sequencing* arrived, and a steep step-change followed. Many cost histories highlight a particularly sharp inflection around January 2008, when these technologies began to displace older methods at scale. The effect was not merely that genomes became cheaper to read; it was that sequencing became something you could *use casually*. You could sequence not only the organism you were studying, but the contaminants in your lab, the mixed population in a fermentation, the microbial community in a pond, the off-target edits in a genome engineering experiment. Sequencing became a way of checking your work.

Writing DNA has undergone its own transformation. Early genetic engineering often meant cutting and splicing whatever DNA you could get your hands on: a gene from one organism, a plasmid from another, stitched together with restriction enzymes like a craft project done under fluorescent lights. It worked, and it launched biotechnology, but it carried the limits of bricolage. You were constrained by what existed and by what you could manipulate with a small set of tools.

Commercial DNA synthesis changed the emotional texture of the field. Instead of cloning a fragment through a chain of finicky steps, you could design a sequence on a computer and order it, as if genes were components. Vendors now sell gene fragments, assembled genes, and even libraries of thousands of variants. These offerings are often described in the language of manufacturing: turnaround time, yield, quality control, price per base. The market is competitive enough that marketing claims themselves become a kind of timestamp. In 2025, for instance, Integrated DNA Technologies announced gene fragments priced as low as $0.07 per base, with rapid turnaround, a figure that would have sounded like a typographical error to many labs a decade earlier. Even allowing for the difference between list prices and real-world project costs, and even allowing for the fact that synthesis is not only about the price of bases but also about error rates and assembly limits, the direction is unmistakable: the alphabet of life is being sold at retail.

These cost collapses do more than save money. They change what it means to plan an experiment. When sequencing is expensive, you avoid it. You infer. You guess. You trust that what you built is what you intended. When sequencing is cheap, you verify. Verification becomes a habit, and habits accumulate into culture.

In biology, culture is not only a social word. It is literal. A successful lab depends on the slow transmission of tacit skills: how to avoid contamination, how to pipette without introducing bubbles, how to interpret a gel that refuses to look clean, how to notice that a centrifuge sounds wrong. Traditionally, these skills were passed through universities and research institutes, where the barriers to entry were admissions, credentials, and institutional access.

Community laboratories complicate that pipeline. They are membership-based spaces where people pay dues, take training courses, share equipment, and propose projects. Some are explicitly educational, others more experimental, and many are a mixture. They borrow governance ideas from hackerspaces, makerspaces, and open-source communities: documentation, peer review, public talks, shared protocols. The difference is that the objects on the bench are alive, and so the social contract has to include biosafety.

One of the most revealing features of the community biology movement is that its ethical language appeared early. DIYbio groups circulated draft codes of ethics by at least 2011, emphasizing transparency, safety, accountability, open access, and peaceful purposes. These statements were partly aspirational and partly defensive. They were a way of telling the world, and telling themselves, that democratization did not have to mean irresponsibility. Since then, community labs have produced handbooks and training materials that look, in miniature, like the bureaucratic scaffolding of institutional research: guidance on biosafety levels, waste handling, responsible conduct, and the special sensitivity around human subjects. The existence of such documents does not guarantee good behavior, but it signals a recognition that biology cannot simply adopt the norms of tinkering with electronics. A miswired circuit usually fails locally. A living system can, in the wrong context, travel.

The most famous public symbol of democratized biological design is iGEM, the International Genetically Engineered Machine competition, founded at MIT in the early 2000s. Teams of students build engineered biological systems and present them in a structured jamboree that mixes science fair, conference, and theatre. iGEM is often discussed as an educational program, but it is also an infrastructure for shared imagination. It taught a generation of students to think of DNA as modular and designable, and it turned laboratory work into something that could be narrated to the public without apology. More quietly, it provided a training ground for safety and ethics discussions in a setting where exuberance is the default. When students are encouraged to build, they must also be taught to ask what building implies.

A historical anecdote from computing helps clarify why this matters. In the 1970s, the Homebrew Computer Club in California gathered hobbyists who could not have afforded mainframes and who were impatient with the idea that computation belonged only to corporations and universities. Out of that culture came not only companies, but a new assumption: a computer could be personal. That assumption reshaped the world. Synthetic biology is not at the same stage, and living systems are not silicon, but there is a family resemblance in the way lowered costs unlock a new social topology. People who meet in shared spaces trade tips, share failures, publish designs, and build reputations outside traditional credentialing.

The tools that enable this are not all glamorous. A reliable PCR machine (a thermocycler) is more important to many small labs than a gene sequencer. Centrifuges, incubators, pipettes, and gel boxes are the workhorses. The magic, if there is magic, lies in how these mundane devices interlock with a supply chain that now includes custom DNA delivered by mail, cloud-based design software, and public repositories of plasmids. A person with modest equipment can perform experiments that once belonged only to well-funded institutions because the missing pieces can be bought as services.

This "biology as a service" model has an interesting psychological effect: it separates design from fabrication. You can design a DNA sequence at midnight, submit it online, and receive a tube of dried DNA days later. The wet-lab labor shifts from making the sequence to integrating and testing it. That changes who participates. Some people are more comfortable in code than at the bench; others are natural experimentalists but not designers. As the interfaces improve, the field begins to resemble software development, with its own versions of libraries, version control, and debugging.

But biology resists the software metaphor at precisely the points where garage culture is most likely to overreach. Cells are not deterministic machines. A genetic circuit that behaves beautifully in one strain may misbehave in another. A promoter that produces a predictable level of expression on Tuesday may drift by Friday because the cell's metabolic state has changed. Even the act of measuring can perturb the system: fluorescent proteins take time to mature; reporters impose metabolic burdens; selection markers bias what survives. Synthetic biologists spend a great deal of time dealing with context, crosstalk, and noise.

That unpredictability is not merely an inconvenience; it is a safety issue. A person who thinks of biology as plug-and-play may underestimate risk. A person who understands how finicky engineered organisms can be may also recognize that the finickiness cuts both ways: many engineered microbes are fragile outside their intended conditions. They do not automatically become invasive threats. The danger is less in the cartoon image of a monster bacterium and more in the serious, prosaic problem of dual-use knowledge and capability: tools and methods that can be used for beneficial projects can also be misused.

So democratization forces a dual argument. On one side, cheaper sequencing and synthesis spread capabilities that can be socially valuable: local environmental monitoring, educational access, low-cost diagnostic ideas, creative biomaterials, and a broader base of people who understand biotechnology well enough to participate in public decisions about it. On the other side, spreading capability widens the set of actors who can attempt harmful or reckless things, whether intentionally or accidentally. The responsible response is not nostalgia for high barriers. It is the construction of norms, training, and layered containment practices that scale with accessibility.

Containment, in practice, often looks like a set of small habits rather than a single dramatic safeguard. Work with non-pathogenic strains. Avoid environmental release. Autoclave waste. Keep clear records of what organisms are present. Learn the difference between biosafety levels and respect their logic. Use genetic "safety locks" where appropriate, such as engineered auxotrophies that require specific nutrients not found outside the lab. None of these habits is glamorous, and that is exactly why they matter. A culture that treats safety as part of craftsmanship tends to produce fewer spectacular failures.

The image of the garage biologist, then, should not be romanticized as solitary brilliance. The more realistic picture is collective competence: people learning together, borrowing rules from professional labs, adapting them to community settings, and sometimes inventing new ones. The garage, in this sense, is not a private room but a network of shared benches connected by cheap DNA, cheaper sequencing, and online knowledge.

A concrete example makes the democratization feel less abstract. Imagine a student group trying to build a biosensor for lead in water. They do not need to sequence a genome from scratch. They can start with a well-characterized bacterial strain, order a designed DNA fragment that couples a lead-responsive regulatory element to a fluorescent reporter, and assemble it into a plasmid using kits that are now standard. Within weeks, they can test whether the bacteria glow in the presence of lead, measure response curves, and troubleshoot problems by sequencing their constructs to confirm what they actually built. They might never touch a high-end sequencer, but they will benefit from the fact that sequencing is cheap enough to be routine somewhere, and from the fact that synthesis is cheap enough to treat DNA as editable text.

The decisive change is not that everyone can do everything. It is that the distance between an idea and a prototype has shortened. In older biotechnology, the bottleneck was often access to specialized equipment and specialized sequences. Now the bottleneck is more likely to be conceptual clarity, experimental skill, and responsible judgment. Those are harder to mass-produce than base pairs.

The garage biologist is sometimes portrayed as a threat, sometimes as a hero, and sometimes as a punchline. The more interesting interpretation is that they are a symptom of a deeper transformation: biology is becoming an engineering medium, and engineering mediums, once cheap enough, attract hobbyists. The first people to show up are rarely the most polished. They are the ones willing to fail in public, to learn by doing, to share half-working protocols, and to argue about standards over cheap pizza.

That messiness can look alarming from the outside. From the inside, it looks like a field learning how to distribute itself without dissolving into chaos. The future of democratized synthetic biology will depend less on any single invention than on whether that distribution can be paired with a shared seriousness: a sense that life is not just another material, and that the power to write DNA, now within reach of so many hands, is inseparable from the obligation to handle it with care.

Chapter 5

Rise of the Xenobots

What happens when skin cells are liberated from their biological duties and instructed by artificial intelligence to build something entirely new? We enter the laboratory where frog tissues are reimagined as programmable organisms, blurring the ancient line between animal and machine and challenging our definition of a 'robot'.

5.1 The Algorithm and the Embryo

When most people hear the word *robot*, they picture an object assembled from parts: metal limbs, motors, batteries, gears. A living robot sounds like a contradiction, as if the category were being stretched for shock value. Yet the trick behind xenobots was not a flashy new material or a secret gene edit. It was a change in *how* a body gets designed: not sketched by a human engineer and then manufactured, but searched for by a computer, tested in a virtual world, and then built from the most ordinary of biological starting points—embryonic cells from a frog.

That change matters because living tissues are not passive building blocks. A steel beam does not decide to curl, migrate, or fuse with its neighbor. A clump of embryonic skin cells does, because it carries its own developmental habits: ways of sticking together, sealing wounds, smoothing surfaces, and responding to mechanical stress.

In an embryo these tendencies are tightly chaperoned by chemical signals and geometry. Remove the cells from that context, and they still retain the underlying competencies, like a team of skilled workers arriving on a job site without blueprints. The xenobot project was, in a deep sense, an experiment in giving such workers a new assignment.

A historical analogy helps. In the nineteenth century, the mathematician and engineer Isambard Kingdom Brunel—famous for bridges, ships, and railways—did not merely calculate structures; he pioneered systems for coordinating labor, materials, and design in an industrial world. What changed engineering was not only better iron and better steam engines, but the emergence of workflows: the separation of design from fabrication, the use of standardized components, the ability to iterate on paper before committing to rivets. Xenobots belong to that tradition of workflow innovation, but with a twist: the components are cells, and the paper prototype is a physics simulation.

The key idea is deceptively simple. Suppose you want a tiny living construct that can move across a surface. You could try to sculpt a shape by intuition and hope it works. Or you could treat the shape as an unknown and ask a computer to search for it. The second approach resembles what nature itself does through evolution: generate many variants, test them, and keep the ones that perform best. An *evolutionary algorithm* is a computational method inspired by that logic. It begins with a population of candidate designs (often random). Each design is evaluated with a *fitness function*, a score that encodes the goal—for example, distance traveled in a fixed time. The best-performing designs are then copied and slightly altered (mutated), sometimes combined (recombined), producing a new population. After many rounds, the population tends to contain designs that accomplish the goal under the rules of the simulation.

The simulation rules are crucial, because the computer is not imagining an abstract creature in a void. It is running a simplified physics engine: gravity, friction, elasticity, and the forces generated by tiny contractile elements that stand in for muscle-like tissue. If the rules are wrong, the algorithm will find designs that work only in a virtual fantasy. If the rules are plausible, the algorithm can become an unusually patient designer, trying thousands or millions of options that no human could sketch, discard, and refine.

This is where the phrase "designed by a supercomputer" can mislead. The computer is not a genius in the human sense. It does not *understand* why a shape works. It performs an enormous number of cheap experiments, guided by a numerical score. The creativity comes from the combination of a large search space, a goal that is stated clearly, and physical rules that keep the search tethered to reality. In the early xenobot work led by Michael Levin and Josh Bongard (2020), the algorithm explored body plans built from small blocks representing tissues with different properties: some passive, some contractile. Many designs were useless. Some spun in place. Some jittered. A few discovered a gait—a way of shifting contact and force against the surface—and those were promoted in the evolutionary process.

What made the results so arresting was their strangeness. Engineers, trained on wheels and legs, have a strong sense of what *should* work. But the algorithm is indifferent to human expectations. It will propose lopsided bodies, asymmetries that look like mistakes, and cavities that seem purposeless. If a bizarre protrusion increases the fitness score by a fraction, it stays. There is a faint echo here of early aviation, when experimenters tried gliders with awkward wings and unstable frames. Only later did flight settle into familiar silhouettes. In xenobots, the familiar silhouettes never arrive, because the scale is so small and the materials so soft that the rules of good design are genuinely different.

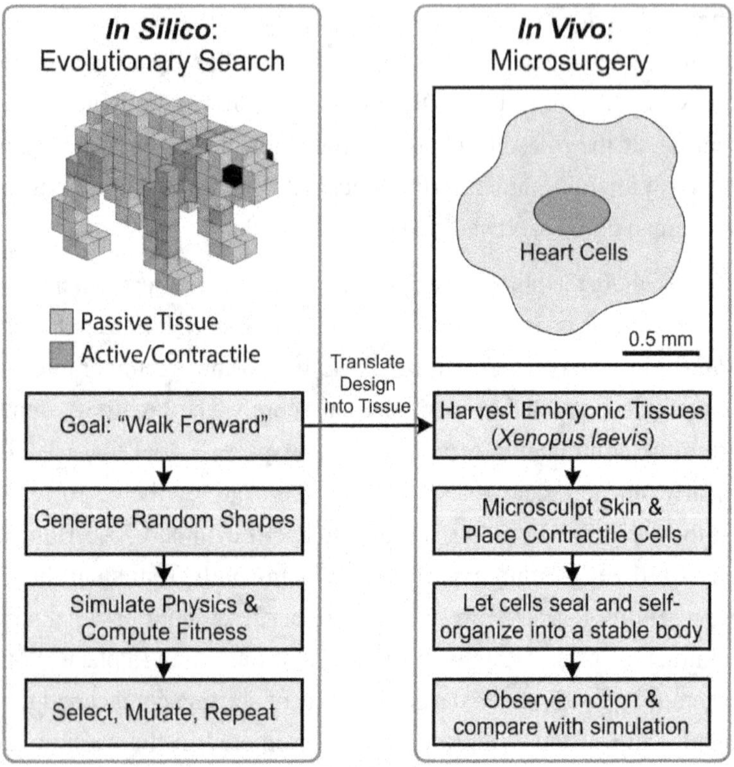

Calling this process "AI" is partly accurate and partly a cultural shorthand. In technical terms, the early pipeline is closer to automated search and optimization than to a system that learns concepts from data. But the broader point stands: the design is not a hand-drawn blueprint. It is an artifact of computation, emerging from a competition among variants inside a simulated environment. That environment is not merely decorative; it is a compressed hypothesis about how living tissue will behave when arranged in a certain geometry and then released to do what it does best: pull, relax, adhere, and remodel.

There is a subtle philosophical shift here, and it can be felt even by non-specialists. Traditional bioengineering often starts with genetics: change DNA, change proteins, change cell behavior, and eventually change tissue. Xenobots start one level up, with *morphology*—the shape and arrangement of tissues—and treat genetics as fixed. The cells are not rewritten; they are reassigned. Skin cells become the outer hull of a tiny moving structure. Heart muscle cells, whose natural job is rhythmic contraction, become an engine. The design question becomes: what arrangement of these existing capabilities yields the desired motion?

One reason *Xenopus laevis* became the workhorse is pragmatic: frog embryos are robust, their development is well studied, and embryonic cells are famously cooperative about forming tissues. Anyone who has watched time-lapse footage of embryonic cells reorganizing can feel the uncanny competence at work. Cut a tissue fragment and it rounds up, like a droplet of viscous fluid seeking the lowest-energy shape. Bring two fragments together and they fuse. Make an incision and the cells crawl to close it. This is not magic; it is the collective result of adhesion molecules, cytoskeletal tension, and biochemical signaling. But from the outside it looks like will.

The wet-lab translation, in the earliest xenobot demonstrations, was performed with microsurgical tools under a microscope. Imagine working at a scale where a millimeter is a long distance, where a tweezer tip is a blunt object, where every move risks tearing tissue. The procedure resembles sculpture more than assembly. A small piece of skin tissue is trimmed into a chosen shape, then a patch of cardiac tissue is positioned so that its contractions can deform the body. Once the tissues are placed together, the most impressive manufacturing step is what happens on its own: the edges seal, the cells knit into a coherent surface, and the construct stabilizes into a compact living object.

This is the embryo's gift to engineering. Embryonic tissues are not like cut foam or molded silicone. They come with repair and finishing processes built in. A human engineer can position components, but the final integration is performed by the cells themselves, executing ancient developmental programs. The xenobot is therefore neither a simple organism nor a conventional machine. It is closer to a *composite behavior*: the outcome of mechanical design choices interacting with cellular self-organization.

One can see, in this, a distant cousin of another famous shift in biological thought: the move from vitalism to mechanism. Vitalists once argued that life required a special essence. Mechanists insisted that life could be understood in terms of physical processes. Xenobots do not resurrect vitalism, but they do complicate naive mechanism. The parts are physical, but they are also agents at their own scale, responding to signals and stresses, changing shape, changing adhesion, sometimes changing the very boundary conditions the engineer thought were fixed. The "machine" is not only the arrangement; it is the ongoing negotiation among living parts.

A concrete example makes this less abstract. The computer may output a design that looks like a squat body with an off-center protrusion. In simulation, this protrusion drags against the ground in a way that biases motion forward. In the lab, the same geometry may produce a slightly different gait because real tissues deform unevenly, or because the heart-cell patch contracts with a particular rhythm. But the overall behavior can still resemble what the simulation predicted: a persistent drift in one direction, driven by rhythmic contractions. The correspondence is never perfect. That imperfect correspondence is not a failure; it is the price of working with living matter and simplified models. The astonishing part is that the correspondence is good enough to make the pipeline useful.

The evolutionary algorithm also encourages a kind of honesty about goals. In nature, evolution optimizes only what affects survival and reproduction in a given environment. In the computer, the fitness function defines what "success" means. That definition can be narrow, and the algorithm will exploit it. Ask for speed, and it may produce a design that races forward briefly but then flips over and stalls. Ask for distance over time, and it may produce something slower but more stable. Ask for carrying a payload, and the shape changes again. This can feel like a parable about bureaucracy: you get what you measure. But it is also a scientific instrument, because it forces the designer to formalize intentions and confront hidden assumptions.

At this point, a question naturally arises: if the algorithm can propose a shape, why not simply 3D-print it? The answer is that a printed object would be inert, and at millimeter scales inert objects struggle. Viscous forces dominate; surfaces are sticky; tiny motors are difficult to integrate. Living tissue solves these problems by being its own actuator and power source, at least for a time. Heart cells contract without wires. Cilia beat without gears. Cells can generate forces relative to their size that engineered micro-actuators still find hard to match. Biology is a mature technology, honed over hundreds of millions of years. The xenobot pipeline borrows that maturity while trying to redirect it.

Another reason the pipeline is scientifically provocative is that it treats morphology as something that can be *searched* rather than *explained*. Developmental biology has traditionally tried to answer why a frog embryo becomes a frog: how genes, signals, and mechanics produce a particular anatomy. Xenobots invert the question: given the rules of tissue behavior, what anatomies are possible, and which ones achieve a goal? That is not a replacement for explanation, but it is a different mode of inquiry. It turns morphogenesis into a design space.

Levin and Bongard's collaboration made that inversion feel natural because it united two intellectual lineages that rarely speak fluently to each other. Developmental biology is rich in qualitative descriptions and mechanistic pathways; evolutionary computation is rich in quantitative optimization and emergent form. When the two are combined, the computer becomes a generator of hypotheses about what living tissues can do when released from their usual developmental narrative.

Public fascination quickly brought a second complication: language. The term "xenobot" sounds like a new creature, and the images—tiny blobs moving under a microscope—trigger instincts about animals. But these constructs do not have neurons, do not have organs, do not reproduce in the ordinary way, and do not behave like any known larva. They are better understood as living tissues configured to produce a behavior, like a biological gadget. The distinction is not pedantic. It shapes ethical reactions. A gadget is something we feel permitted to redesign freely. An animal is something we feel obligated to treat with care. Xenobots sit uneasily between those moral intuitions.

That unease has practical consequences even at this early stage, because the pipeline can be generalized. Once you have a way to translate computational designs into living assemblies, the question becomes not only what you can *build*, but what you should *allow*. An evolutionary algorithm is indifferent to intent. If the goal were stated badly, the resulting forms could be fragile, unpredictable, or difficult to contain. Early xenobots were confined to lab dishes and survived for limited periods on internal energy reserves. That limitation is a form of safety by default, a biological battery that runs down. But safety cannot be left to default settings when the methods become more powerful and the applications more ambitious.

Containment, in the broad sense, means making sure a living machine stays where it is supposed to stay and does what it is supposed to do. For constructs that cannot survive outside a controlled environment, containment is straightforward. For constructs that might one day operate in soil, water, or the body, containment becomes a design requirement. Here the algorithmic workflow can help rather than hurt: constraints can be encoded. One can demand that a design requires a particular nutrient absent in nature, or that it degrades after a set time, or that it remains inactive unless triggered by a specific cue. Those are not mere afterthoughts; they can be made part of the fitness function, just as stability and speed are.

There is also a softer kind of containment: conceptual containment. The public story of xenobots sometimes drifted toward sensationalism, as if a new form of life had been created from scratch. In reality, the achievement was more subtle and, in some ways, more powerful. The genome did not need to be rewritten to produce novelty. Novelty emerged from rearrangement, from asking what familiar cells might do when placed in unfamiliar geometries and then evaluated under unfamiliar criteria. That suggests a broader lesson about biology: a great deal of behavior is not locked in DNA like a script, but arises from interactions among cells, their physical environment, and the signals they exchange.

The embryo is the quiet protagonist of this story. Embryos are not only collections of cells; they are systems for generating organized complexity, for turning local rules into global form. The xenobot pipeline treats embryonic tissue as a programmable medium, but the "programming" is not a line of code inserted into DNA. It is a combination of shape, boundary conditions, and selection. The algorithm proposes, the microsurgeon arranges, and the cells complete the construction by following their own rules.

Watching a xenobot move is therefore like watching a compromise between two kinds of design. One is explicit and external: the computer's chosen geometry, the human's careful assembly. The other is implicit and internal: the tissue's drive to round up, to seal, to contract rhythmically, to preserve integrity. The motion emerges from their intersection. It is neither fully engineered nor fully natural, but something that exists because computation and embryology have begun to share a common language: performance.

And once performance becomes a shared language, the relationship between biology and engineering changes. The question stops being whether living matter can be made to behave like a machine. It becomes whether living matter, searched and shaped by algorithms, can reveal behaviors that engineering would never have imagined to ask for, let alone to draw by hand.

5.2 Flesh and Frame

A xenobot built entirely from frog cells is a provocation: a reminder that cells, when cut loose from their usual developmental script, can be arranged into a body that performs a task. Yet pure tissue has limitations that become obvious as soon as one tries to do anything precise. A clump of cells can crawl or glide, but can it swim in a chosen direction? Can it carry a sensor, push fluid through a microchannel, or squeeze through a narrowing gap without tearing? Living tissue is strong for its size, but it is also soft, variable, and eager to remodel itself. Engineers who fall in love with biology quickly discover the second half of the romance: you do not merely *use* living matter; you negotiate with it.

Biohybrid soft robotics begins at that negotiation. The basic premise is to split the problem into two complementary parts. The *frame* provides geometry, anchoring points, and predictable mechanics. The *flesh* provides actuation: the ability to bend, pulse, contract, and recover—often more efficiently and gently than any tiny motor could manage at the same scale. Put differently, the synthetic components carry the burden of reliability, while the living components carry the burden of movement. When the match is good, the result feels less like a robot with a biological decoration and more like a new category of machine: one that can be driven by physiology.

The story has older roots than its current vocabulary suggests. In 1959, Richard Feynman delivered his famous lecture "There's Plenty of Room at the Bottom," imagining miniature machines and the strange engineering rules that would apply as objects became microscopic. He could not have predicted optogenetics—using light to control cells—nor the modern toolkit of microfabrication. But he understood the central point: when you shrink machines, new forces dominate, new materials become useful, and the line between living and manufactured systems begins to blur. Biohybrid robots are one way that blur has become tangible.

At millimetre and sub-millimetre scales, a traditional motor is awkward. It needs wires, power, and rigid components that are hard to miniaturize without friction and wear devouring the output. Muscle tissue, by contrast, is a microscopic engine that comes pre-assembled. Muscle cells convert chemical energy into mechanical work, and they do it repeatedly. They can also be grown into specific shapes, aligned so that many small contractions add up to a useful pull. In the lab, muscle behaves like a living material that wants to be organized: if it is given a scaffold with grooves or patterned chemistry, it tends to align its fibres along those cues, forming something that resembles a tiny strap.

A concrete illustration of the biohybrid approach is the stingray-inspired swimmer created by Kevin Kit Parker's group (2016). The design borrowed a simple trick from marine animals: use a flexible body and rhythmic waves to push against water. The "skeleton" was not bone but an engineered, flexible structure. The "muscle" was made from cultured rat heart muscle cells, patterned so their contractions produced a coordinated bend rather than a useless twitch. The result moved through fluid with a grace that felt unsettling precisely because it was not a simulation. You could watch it under a microscope and recognize, in miniature, a biological style of motion: smooth, compliant, and hard to replicate with rigid machines.

The stingray is also a useful cautionary tale about what biohybrid engineering demands. Muscle cells do not contract on command unless they are wired into a control system. Heart cells, in particular, have their own rhythms. Left alone they beat, and the beat is not easily switched off. That is a feature if you want a constant pulsing motion, but a problem if you want steering. The solution in many biohybrid devices is to introduce a *control interface*: a way to translate an external signal into a cellular response.

Two interfaces dominate because they match the scale and delicacy of the material. The first is electricity. Muscle cells are excitable; a brief electrical pulse can trigger contraction. By placing electrodes in the environment or embedding conductive features in the frame, engineers can pace the tissue, speeding it up, slowing it down, or activating one side more than the other. The second interface is light, especially when combined with optogenetics. Optogenetics is a technique in which cells are genetically modified to express light-sensitive proteins so that illumination can trigger electrical activity and, downstream, contraction. Light has an obvious appeal: it can be patterned in space and time without physical contact, like puppeteering at a distance.

The elegance of these interfaces can make the engineering look cleaner than it really is. Electrical pacing requires careful control of pulse strength and timing; too little does nothing, too much damages cells. Optogenetics adds another layer: the biology must be modified so the cells respond to light, and the light must reach the cells without heating them or degrading the tissue. Both approaches turn a simple question-*how do we make it move?*-into a coupled design problem spanning physiology, materials science, and control theory.

Soft robotics provides the conceptual bridge. A soft robot does not rely on stiff joints. It changes shape through deformation, often distributing forces across its body. This makes it gentle, adaptive, and surprisingly resilient to damage. Living tissues are the ultimate soft actuators: they deform naturally and can survive bending that would snap a rigid linkage. But their softness also means their behavior is strongly shaped by the frame they are attached to. A small change in stiffness at an anchoring point can redirect the line of force; a slight asymmetry in thickness can introduce an unintended twist. In biohybrid devices, geometry is not cosmetic. It is a form of programming.

This idea-geometry as programming-connects directly to the xenobot mindset. When frog tissues were sculpted into new shapes, the shape changed how contractions translated into motion. Biohybrid devices take the same principle and make it more predictable by adding a synthetic scaffold. If the tissue is the engine, the scaffold is the transmission.

Engineers build those transmissions in ways that can sound almost too simple: thin polymer films that bend, hydrogels that swell, micropatterned surfaces that guide cell alignment. Yet simple components can produce complex movement when arranged cleverly. Consider a polymer strip with a muscle band attached on one side. When the muscle contracts, the strip bends toward that side. Release the contraction, and it springs back. Repeat, and you get a flapping motion. Put two such strips in a V-shape, and you get a swimmer. Add asymmetry in stiffness, and you get a turn. A great deal of biohybrid robotics is the art of turning a one-dimensional contraction into a two- or three-dimensional motion.

The appeal of a synthetic frame is not only control but also integration. Sensors, channels, and cargo compartments are difficult to build out of living tissue alone. A frame can carry microelectronics, fluorescent markers, magnetic particles, or microfluidic pathways that let the device pump or sample fluids. In principle, a biohybrid robot could be both a mover and a measurer: a tiny system that navigates a chemical gradient and records what it encountered. The more the frame can standardize the platform, the easier it becomes to compare designs, reproduce results, and scale manufacturing.

That word-*platform*-signals a shift in ambition. The earliest xenobots were made by hand, a triumph of proof-of-principle but not of throughput. Subsequent work, including the more self-organizing xenobot platform described by Blackiston and colleagues in *Science Robotics* (2021), pushed toward constructs that form more readily and can be produced in larger numbers. A key move was to exploit cilia: tiny hairlike structures on cell surfaces that beat rhythmically and can propel fluids or the cells themselves. Cilia are familiar from biology textbooks as the machinery that moves mucus in airways or helps move eggs in reproductive tracts. On the surface of a tiny construct, coordinated cilia beating can generate a steady glide without needing an internal patch of heart muscle at all.

Cilia-based locomotion is a beautiful example of biological repurposing. The same cellular apparatus that, in an embryo or organ, serves a specific physiological function can become a propulsion system when placed in a different geometry. If heart muscle offered a kind of biological piston, cilia offer something closer to oars. They also change the control problem. Cilia beating is often more uniform and continuous, which can yield smoother motion. It can also be influenced by the patterning of tissue: where ciliated cells end up on the surface, how they are oriented, and how the surface curvature channels the resulting flow.

The *frame* in a cilia-driven construct may be less visible, but it is still there in the form of boundary conditions. Even without an obvious synthetic scaffold, the geometry imposed during formation acts as a frame: a way of corralling self-organization into a useful shape. The distinction between a pure xenobot and a biohybrid device is therefore not always binary. Many systems lie along a continuum from mostly living to heavily scaffolded. What matters is the design philosophy: to treat living matter as an active component that can be guided by engineered constraints.

One of the most striking demonstrations in the 2021 *Science Robotics* work was not locomotion but memory. Memory, here, does not mean cognition. It means the ability to record an event in a durable way-an engineering need as old as the first data logger. The researchers demonstrated a proof-of-principle "writable" molecular memory by using a photoconvertible protein: a molecule that changes its fluorescent color when exposed to a particular wavelength of light. Shine the right light, and the protein shifts from one state to another, leaving a visible trace that can be read later under a microscope. The construct becomes, in effect, a living device that can be marked by experience.

This kind of memory is modest compared to neurons and brains, but it matters because it shows how quickly the space of "robot functions" can expand once biology is part of the toolkit. A conventional micro-robot might store information in electronics, but electronics are hard to embed in soft, wet, living matter without encapsulation. A molecular recorder is naturally compatible: it is part of the cellular interior. It suggests a direction in which biohybrid systems do not merely *move* but also *sense and register*—using biological chemistry rather than silicon circuits.

The public imagination often jumps from such demonstrations to medical fantasies: tiny living machines patrolling the bloodstream, seeking tumors, repairing tissues. Those visions are not impossible, but biohybrid engineering forces a more grounded view of what it takes to operate inside the body. The body is not a benign aquarium. It is warm, crowded, chemically complex, and immunologically vigilant. Any device meant to function inside it must contend with immune recognition, clearance mechanisms, and the risks of unintended interactions. Living components can help here-they may be more biocompatible, more adaptable-but they also introduce new uncertainties, such as how long they survive, what they secrete, and how they change over time.

This is one reason many of the most realistic near-term applications for biohybrid soft robots are not free-roaming medical agents but controlled laboratory and industrial tasks. A biohybrid pump that moves fluid through a chip is easier to manage than a biohybrid swimmer in a bloodstream. A soft gripper that gently handles fragile tissue samples in a dish is easier to validate than a device that must navigate a living organ. In these settings, the advantages of living actuators-gentle force, self-repair tendencies, high efficiency at small scales-can be exploited without taking on the full complexity of an organismal environment.

Reliability is the recurring engineering headache. Muscle tissue changes as it matures. It responds to nutrient availability and oxygen levels. It can fatigue. It can remodel itself in response to mechanical load, altering performance over time. Engineers are used to components with datasheets. Cells come with biographies. Two batches of ostensibly identical cells can behave differently, because biology is noisy and history-dependent. Biohybrid robotics therefore borrows techniques from tissue engineering: standardized culture conditions, careful characterization, and, increasingly, automation to reduce variability introduced by human handling.

The frame also has a second job beyond mechanics: it can help keep living tissue alive. Diffusion limits become important at small scales. Cells need oxygen and nutrients; waste products must be removed. A frame can incorporate channels to perfuse media, or porous materials that improve exchange. In some designs, the frame is less like a skeleton and more like an organ support system, ensuring that the biological actuator does not simply starve.

Ethical questions hover nearby, though they are often more about framing than about suffering. Cultured muscle tissue does not have a nervous system. It does not experience pain in any recognizable sense. Yet the use of animal-derived cells, especially from mammals, raises concerns about sourcing and commodification. When optogenetics is used, genetic modification enters the picture, which triggers additional public sensitivity. A responsible account should avoid both complacency and melodrama. Biohybrid devices are not sentient machines, but they are part of a broader trend: treating living tissue as an engineering substrate. That trend deserves scrutiny because it alters what laboratories, companies, and societies consider normal.

There is also a security dimension that appears, at first glance, remote. A biohybrid soft robot is not a pathogen. It does not replicate in the usual way, and it is typically dependent on lab conditions. Yet any technology that makes biological construction easier can be misused in principle. The most important response is not to inflate the threat but to normalize good governance: clear reporting, containment practices, and transparency about what systems can and cannot do. In the short term, the most likely harm is not deliberate misuse but accidental hype: the claim that a twitching piece of tissue is an autonomous agent, or that a proof-of-principle memory marker is a mind. Such claims corrode trust and make genuine progress harder to discuss.

Under the microscope, the most compelling fact about biohybrid robots is mundane: they work because life is already a master of soft actuation. Engineers are not inventing a new kind of muscle; they are borrowing an existing one and wrapping it in constraints until it behaves as part of a machine. The constraints can be geometric, electrical, optical, or chemical. Each constraint is a handle, and each handle brings both power and responsibility. Pull too hard, and the system fails; pull with care, and living matter reveals capabilities that rigid machines struggle to match.

A piece of living muscle contracting on a polymer skeleton can look like a toy, until one notices the scale of the forces involved, the precision of the patterning, the stability of the rhythm. At that point the spectacle becomes less about novelty and more about a pragmatic possibility: machines that are not assembled from lifeless parts but grown and guided, blending the predictability of engineered frames with the restless competence of cells.

5.3 Rewiring the Biological Software

A living machine is not only a piece of moving tissue. It is a piece of tissue that *knows how to become itself*, and that knowledge is distributed through the cells like a set of habits. Cut a fragment from an embryo and it rounds up, sealing its surface. Scratch it and cells crawl to close the gap. These are not reflexes in the neurological sense. They are competencies built into the collective: decisions made without a brain, coordinated by signals that are older than animals themselves.

Most popular accounts of biology begin and end with genes, as if DNA were a dictator issuing orders. Genes matter enormously, but they are not the whole story of form. Two cells with the same genome can become skin or nerve; the difference lies in what they *express* and how they *communicate*. Chemical signals-hormones, growth factors-are familiar messengers. Less familiar, and in many ways more surprising, is that cells also talk in electricity.

Bioelectricity is easy to misunderstand because it carries the glamour of neurons and the brain. That is the flashy version: electrical impulses as thoughts and sensations. The quieter version is older and more widespread. Every living cell maintains an electrical potential across its membrane, a small voltage created by pumps and channels that move ions-charged atoms-such as sodium, potassium, chloride, and calcium. This voltage is called the *transmembrane potential*, often abbreviated as V_{mem}. If that phrase sounds technical, it helps to translate it into an image. Think of each cell as a tiny battery with a living casing. It is constantly spending energy to keep one side of the membrane more charged than the other, and that stored difference is a form of information.

Cells are not isolated batteries. They connect to neighbors through structures called *gap junctions*, which allow ions and small molecules to pass directly from cell to cell. A tissue can therefore form an electrical network, not unlike a soft, shifting circuit board. Instead of copper wires and solder, it uses membranes and channels. Instead of a fixed schematic, it has a dynamic architecture that can change as cells move, divide, and differentiate. In that network, voltage patterns can spread, stabilize, and interact with chemical signals.

Michael Levin and collaborators have been among the most persistent advocates for treating these voltage patterns as a kind of biological control layer-a "software" that helps specify anatomy. The metaphor is imperfect, but it is useful if handled carefully. Software, in an ordinary computer, is an abstract pattern that tells hardware what to do. In tissues, bioelectric patterns are not abstract; they are physical states. Yet they can function like instructions because they influence gene expression, cell migration, proliferation, and the timing of differentiation. Change the voltage landscape and, under the right conditions, you can change what the tissue becomes.

A historical anecdote makes the idea less abstract. In the late eighteenth century, Luigi Galvani famously observed that frog legs could twitch when touched by metal, and he concluded that animals possessed an intrinsic electricity. His rival Alessandro Volta argued that the metals generated the electricity, leading to the invention of the battery. The debate is often told as a quaint prelude to modern electrophysiology, but it contains a deeper moral: biological electricity is real, and it is easily confused with the electricity we impose from outside. Two centuries later, the bioelectric view of morphogenesis asks a similarly subtle question. Are voltage changes merely byproducts of development, or are they instructive signals that development uses?

The modern answer is that they can be both. Development is a web of feedback loops. Mechanical stress changes ion channel activity; voltage affects gene regulation; gene products alter the cytoskeleton; the cytoskeleton changes tissue tension, and so on. In such a system, there is rarely a single master control. What matters is that voltage is not a passive readout. It is one of the levers the system can pull.

This becomes concrete when a tissue is injured. A wound is not only a hole; it is a disruption of an electrical boundary. The skin, for instance, maintains voltage differences between the inside and outside of the body. When the surface is broken, electrical gradients appear around the wound edge, and cells migrate along those gradients. This phenomenon, sometimes called *electrotaxis* or galvanotaxis, has been studied for decades. It is one reason electrical stimulation has long been explored-sometimes clumsily, sometimes promisingly-in wound healing and regenerative medicine. The broader implication is straightforward: if tissues use electrical cues to coordinate repair, then electrical cues are part of their internal language for maintaining form.

Xenobots are compelling in this context because they tease apart the usual relationship between tissue and goal. In an embryo, bioelectric patterns are in service of making a frog. In a xenobot, the same cells are arranged so their collective behaviors produce locomotion or other task-like outcomes. The cells have not been genetically edited into a new species. They are running familiar cellular programs in an unfamiliar geometry. That unfamiliar geometry changes their interactions, including electrical ones. The "software" is not overwritten by a new genome; it is redirected by new boundaries and constraints.

The shift becomes especially vivid in xenobots whose motion is driven by cilia. Cilia appear in many tissues through ordinary developmental patterning. On a frog's skin, ciliated cells help move fluid. In a xenobot, the distribution and orientation of cilia can become a propulsion system. The same cellular hardware is being used, but the collective outcome depends on how the cells organize themselves on a small 3D surface. If one wants to treat a xenobot as a device, the problem is not only how to build it but how to nudge the tissue into adopting the right internal organization-where cilia are, which way they beat, how stable that arrangement is.

This is where bioelectric thinking becomes more than an interesting side note. If voltage patterns influence cell differentiation and tissue polarity, then they offer a route for steering self-organization without micromanaging every cell. Instead of sculpting the final form directly, one can try to set initial conditions-chemical and electrical-and let the tissue do the hard work. It is a very different style of control from mechanical engineering. The closest analogy might be gardening: you influence growth by shaping the environment, not by placing every leaf.

Levin and colleagues have argued, in reviews around 2018 and 2019, that bioelectric networks provide tissues with a kind of memory and goal-directedness: the ability to restore a normal pattern after perturbation, or to maintain a target anatomy despite noise. "Goal-directedness" can sound mystical, so it is worth grounding it. Consider a thermostat. It "wants" to keep a room at a set temperature, but that desire is merely a feedback loop. In regeneration, something similar may occur at the level of anatomy. Cells behave as if they are correcting deviations from a target pattern: closing wounds, rebuilding missing parts, remodeling toward a stable form. Bioelectric states could be part of the internal representation of that target, a distributed set of signals that guide cells toward a particular outcome.

A striking feature of this view is that it shifts attention from individual molecules to collective dynamics. In gene-centric narratives, to change form you change the DNA or the proteins it encodes. In a bioelectric narrative, to change form you may be able to change how cells are electrically coupled, or how channels are gated, thereby altering the tissue-level pattern of voltage. Because channel activity can be influenced by drugs, ions in the environment, mechanical forces, and light (in optogenetic setups), it opens a toolkit that looks more like "reprogramming" than like editing.

The promise comes with a warning: the same lever that can guide repair can also produce unexpected outcomes. Electrical networks are complex. Small changes can propagate. In a soft, living circuit, local perturbations can lead to large-scale rearrangements. This is part of what makes the field scientifically exciting and clinically challenging. A drug that modifies ion channels may change not only a targeted tissue but also heart rhythm, neuronal excitability, and muscle contraction. Engineering with bioelectricity therefore has to be as attentive to side effects as pharmacology is.

The most dramatic demonstration of what collective dynamics can do-without genetic engineering-arrived with the xenobot work on *kinematic self-replication* by Kriegman, Blackiston, and colleagues (2021). "Kinematic" here means replication by motion. In their experiments, xenobots placed in a dish with dissociated frog cells could corral those loose cells into piles. Under the right conditions, those piles could develop into new motile constructs that, in turn, could participate in the same corral-and-pile behavior. The replication did not occur by a xenobot growing and dividing like a bacterium. It occurred by a moving collective reorganizing other cells into a new collective.

It is hard to encounter this idea without a flicker of unease. Replication is the property we most associate with life's autonomy, and autonomy triggers ethical and existential instincts. The careful, unglamorous reality matters here. The replication was limited and depended on a controlled environment and a supply of suitable cells. It unfolded over days, not seconds, and it did not produce an open-ended evolutionary explosion. Yet the principle is profound: a multicellular arrangement can reproduce not by encoding a blueprint in DNA, but by using geometry and movement to induce other cells to assemble.

From a bioelectric perspective, that replication depends on more than mechanical pushing. The piled cells must self-organize into a coherent, ciliated structure. For that to happen, the cells must coordinate their differentiation and polarity, processes in which bioelectric coupling is thought to play a role. The replication is therefore not simply the reproduction of shape; it is the reproduction of a *developmental process* in miniature. A xenobot becomes a seed of organization, not by carrying genes, but by creating conditions under which other cells execute a familiar self-assembly routine.

The language of "software" begins to earn its keep here. Software can be copied without copying hardware; you can transfer a pattern that causes a system to behave in a certain way. Kinematic replication suggests a biological analogue: behaviors and forms can propagate by inducing self-organization in available matter. If that sounds like science fiction, it may help to remember that embryos themselves are, in a sense, self-organizing matter guided by distributed signals. Xenobots show that those signals can be harnessed in new contexts.

Anthrobots widen the circle further, and they do so in a way that is emotionally and ethically charged. Frog cells are distant enough from humans to feel like a safe playground. Human cells are not. In 2023, researchers reported motile "anthrobots" that self-construct from adult human somatic cells, forming spheroidal multicellular structures that can move using cilia. The work emphasized microenvironmental cues-conditions in the dish and the physical context-as drivers of self-assembly, rather than genetic engineering. One reported observation, in vitro, was that anthrobots moving across a scratched sheet of cultured neural cells were associated with improved closure of the scratch. This is not the same as healing a brain injury, but it is the kind of result that inevitably tugs at translational hopes.

By 2025, further characterization of anthrobots described a life cycle with substantial gene-expression remodeling relative to the starting cells, including activation of genes associated with embryonic patterning. Reports also suggested an epigenetic "age" shift upon morphogenesis, as if assembling into a new multicellular configuration partially reset aspects of cellular state. These findings are still young and deserve cautious interpretation, but they point to a theme that bioelectric and morphogenetic thinking has long suggested: cells carry latent programs for organization that can be awakened or redirected by context.

Here bioelectricity is not a decorative detail. Cilia formation, polarization, and coordinated beating are deeply linked to cellular organization. Voltage gradients can influence cell behavior and can interact with the pathways that establish polarity. If adult human cells can be coaxed into forming motile, organized spheroids, then the control knobs include not only chemical growth factors but also physical and electrical boundary conditions. The "programming language of shape" is therefore not a single language. It is more like a multilingual conversation among genes, mechanics, and bioelectric states.

A skeptic might object that calling this "programming" is an overreach. Programming implies intentionality and precision; biology is messy. That objection is healthy, because it prevents metaphors from becoming dogma. Yet it is also true that engineers routinely program messy systems by exploiting constraints and feedback. The flight controller of an airplane does not calculate every air molecule; it relies on aerodynamic stability and feedback loops. Likewise, guiding a living construct does not require specifying every molecule; it can, in principle, rely on tissue-level attractors-stable patterns that development tends to fall into.

This is why the most interesting form of control in living machines is often indirect. Rather than commanding a contraction, one can bias the tissue toward a pattern that will produce useful behavior. Rather than repairing a cut by stitching, one can encourage cells to migrate and seal. Rather than building a sensor from scratch, one can use a molecular state change as a memory. Each move respects a central fact: living matter is not merely material. It is *active*, and its activity is coordinated.

The ethical resonance becomes sharper once one recognizes that "rewiring" is not only a technical maneuver but also a conceptual one. If cells can be guided into new organizations without changing their DNA, then the boundary between therapy and fabrication becomes less obvious. Regenerative medicine aims to help tissues restore their normal form. Bio-robotics aims to create new forms that perform tasks. The same tools-ion-channel modulators, electrical stimulation, optogenetic control, engineered microenvironments-can serve both aims. In such a landscape, the most responsible stance is neither fear nor exuberance, but clarity: about what is being changed, what is being measured, and what counts as success.

In a dish, a xenobot's self-healing can look almost tender. Slice it, and the cells pull together and smooth the wound edge, sometimes restoring a coherent body. No part has been replaced with a spare. No technician has tightened a screw. Repair is simply what living tissue does when it is given a chance. That simple fact-repair as a default behavior-may be the most practical meaning of "biological software." The software is not a line of code; it is the tissue's stored tendency to maintain integrity, to coordinate across scales, and to treat damage as a solvable problem.

If machines built from metal embody the logic of parts, living machines embody the logic of patterns. Voltage gradients, chemical fields, and mechanical stresses form a layered control system that can be nudged rather than rebuilt. The more that engineers learn to speak those languages, the more "design" begins to look less like assembly and more like persuasion: setting the conditions under which cells choose, collectively, to become something new.

Chapter 6

Defenders of the Biosphere

Nature is often viewed as a fragile victim of industrial progress, yet synthetic biology is transforming it into an active defender. By rewriting the genetic code, scientists are deploying microscopic agents capable of hunting cancer cells within the bloodstream and dismantling toxins in the soil, effectively upgrading the immune systems of both our bodies and the planet.

6.1 The Programmable Pill

Aspirin dissolves. Insulin diffuses. Antibiotics surge through blood, hoping they arrive where they are needed before the liver, the kidneys, and simple dilution blunt their force. For most of modern medicine, a drug has been a substance: a molecule designed to survive the stomach, cross a membrane, latch onto a receptor, and then be cleared. The ambition of a *programmable pill* is different. Instead of swallowing chemistry, you swallow capability: a tiny, living agent that can sense its surroundings, make a decision, and act—not once, but repeatedly, while the situation lasts.

The idea sounds like science fiction because it borrows the grammar of robotics. A microrobot has sensors, a controller, and an actuator. In a living therapeutic, those roles are played by biology itself. Sensors are proteins that bind to a signal and change shape. The controller is a gene circuit: DNA sequences and regulatory parts arranged so that one molecular event can switch another on or off. The actuator might be an enzyme that destroys a harmful metabolite, a protein that recruits immune attention, or a drug payload released only under certain conditions. The organism—often a bacterium with a long history in the human gut, such as *E. coli* Nissle—is both chassis and factory, moving, dividing, and using the body's own chemical gradients as a map.

The metaphor that helps most is the one engineers reach for first: the logic gate. An electronic *AND* gate outputs a signal only when two inputs are present at once. Biology can do that too. A bacterium can be built so that it releases a toxin, or a therapeutic protein, only when it detects both a hallmark of diseased tissue and a second cue that narrows the target further. This is not a poetic analogy; it is a practical strategy for safety. The more conditions you require before a microbe acts, the less likely it is to act in the wrong place.

The appeal becomes obvious in cancer, where the central problem is specificity. Tumours are not foreign objects. They are the body's own cells with altered priorities, hoarding resources, bending blood vessels to their will, and turning immune patrols away at the door. Yet tumours do have characteristic microenvironments. Many are more acidic than healthy tissue because fast-growing cells produce and export lactic acid. Many have regions of low oxygen (hypoxia) because their blood supply is chaotic. Neither acidity nor hypoxia is unique to tumours, but together they begin to look like a fingerprint. A programmable bacterium can be designed to treat that fingerprint as permission.

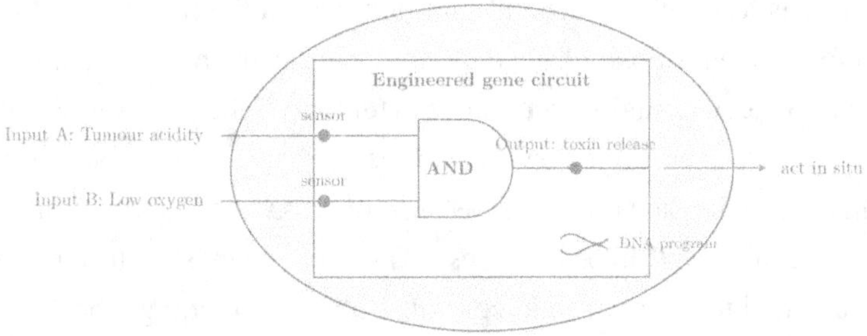

What makes this more than a diagram is the way cells naturally couple sensing to action. In electronics, a sensor is a separate component; in bacteria, sensing is often a transcription factor, a protein that binds DNA and changes whether a gene is read out into RNA. If a transcription factor responds to acidity, it can be wired upstream of a therapeutic gene so that acidity becomes a switch. A second sensor responds to oxygen level. To create an *AND* gate, the bacterium is engineered so that neither sensor alone can turn on the final output. Only when both are triggered does the cell assemble the full set of molecular permissions needed for expression. Biology offers several ways to do this: splitting an activator into two halves that must be produced separately and then rejoin; placing an essential step under the control of one sensor and the final step under another; or building cascades that require two independent keys to unlock the same door.

The first human reaction to such a proposal is usually a fair one: why bring a living thing into the body when conventional drugs already carry risk? The answer is that the body is itself a living, dynamic environment, and the hardest therapeutic problems are dynamic. Inflammatory bowel disease, for instance, is not a fixed lesion but a shifting geography of irritation, repair, microbial imbalance, and immune overreaction. A static pill is like a single broadcast message. A living therapeutic can behave more like a local negotiator, listening first, then responding, and stopping when the conversation ends.

There is also an old medical intuition behind the new tools. Long before anyone spoke of gene circuits, physicians noticed that some bacteria preferentially accumulate in tumours. The observation dates back more than a century in various forms: tumours can be permissive niches for microbes because of leaky vasculature, immune suppression, and hypoxic cores. The modern twist is to take that accidental tropism and add control. Rather than letting a bacterium merely *be* in the tumour, engineers try to make it *behave* there in a constrained, predictable way.

The gut is a particularly attractive arena because it is already a microbial ecosystem and because it offers a practical route in: oral delivery. A bacterium built to live in the intestine does not need to cross into the bloodstream to have clinically meaningful effects. Phenylketonuria (PKU), a genetic disorder of metabolism, provides a striking example. People with PKU cannot properly break down phenylalanine, an amino acid found in many proteins. When phenylalanine accumulates, it can damage the developing brain, and lifelong dietary restriction becomes a form of medicine. An engineered microbe can be designed to metabolize phenylalanine (or a related compound) directly in the gut lumen, reducing the amount that reaches the bloodstream. The therapeutic action is not a single dose of a molecule but an ongoing enzymatic service performed where the substrate appears. In early clinical studies, engineered *E. coli* Nissle strains have been tested as orally administered living therapeutics for PKU, with close attention to safety and how quickly the organisms clear after dosing. The details matter because they reveal what the programmable pill must become to earn trust: not merely effective, but measurable, controllable, and, when necessary, stoppable.

Stoppability is not a rhetorical flourish; it is an engineering requirement. A bacterium divides. That is its defining power and its defining risk. The elegant version of a programmable pill is one that persists just long enough to do the job and then is removed by design rather than by hope. Several strategies exist. One is *auxotrophy*: the microbe is engineered so it cannot synthesize an essential molecule and must obtain it from its environment. If that molecule is supplied in a capsule coating or in a co-administered supplement, the microbe can survive during treatment but dies off when the supply stops. Another strategy is the *kill switch*, a genetic circuit that triggers self-destruction under defined conditions. A kill switch can be tied to the absence of a lab-provided molecule, to an externally administered small molecule drug, or to environmental cues that indicate the bacterium has left the intended niche. The best designs are redundant, because evolution is relentless: a single point of failure will eventually be found.

Evolution is the quiet adversary of all living therapeutics. In a chemical drug, the active ingredient does not mutate. In a cell, every division is an opportunity for a genetic circuit to degrade if it imposes a cost. A gene circuit that forces a bacterium to produce a therapeutic protein might slow its growth; the mutant that disables the circuit can outcompete its dutiful siblings. This is not malice. It is arithmetic. For a programmable pill to work over days or weeks, the program must remain intact in the face of selection.

A clever response to this problem is to design circuits that *remember* rather than continually express. If a bacterium needs only to report that inflammation occurred, it can be engineered to flip a durable genetic switch when it detects an inflammatory marker, and then stay in that state even after the signal disappears. That way, the therapeutic or diagnostic information can be recovered later, for example from a stool sample, without requiring constant high-level gene expression in the gut. In 2017, a memorable demonstration showed engineered bacteria functioning as long-term living diagnostics of gut inflammation in mice by using such memory architectures, confronting precisely the issue of stability over extended residence. The point was not merely that bacteria can sense inflammation, but that they can store the fact of sensing in a form that survives the noise of a living intestine.

Once you accept that a bacterium can *store* a medical event, the next question is how to *read* it. Traditional diagnostics often demand a clinic, trained personnel, and expensive machines. Living diagnostics tempt a different workflow: the body becomes the incubator; the output is designed to be accessible. One early example from Tal Danino and colleagues (2015) illustrated the use of engineered microbes to detect tumours and route a signal into urine, turning an internal event into an external sample that is familiar to medicine. The elegance is in the interface. Urine is already part of diagnostic culture; it can be collected noninvasively, and it does not require disturbing the gut ecosystem. In these schemes, engineered bacteria perform the hard work of locating or sensing the target, while the readout is a simple assay of a body fluid.

This interface problem is also why cell-free biology has become part of the programmable pill story, even though it removes the living organism entirely. Cell-free systems take the molecular machinery of life—the enzymes and ribosomes that read genetic instructions and make proteins—and extract it from cells into a prepared mixture. Freeze-dried onto paper or other porous materials, the mixture can be shelf-stable, transported cheaply, and activated by adding a drop of water containing the sample. The result is a diagnostic platform that behaves like a living circuit without the complexities of keeping an organism alive. In practice, this means that genetic logic can be deployed in places where cold chains are unreliable and laboratories are scarce. A paper-based test can be programmed to produce a color change in response to a pathogen's DNA or a toxin, using the same conceptual toolkit as the bacterial *AND* gate but in a safer container.

The relationship between living therapeutics and cell-free diagnostics is not competition so much as division of labour. Living systems excel at operating in complex environments: the gut, a tumour core, mucus, the crowded chemistry of inflammation. Cell-free systems excel at standardization: they are easier to regulate, easier to store, and easier to distribute at scale. Together they sketch a future in which a programmable pill might be paired with a programmable test: one to act, one to verify.

Verification is not a luxury. If a microbe is going to make decisions inside a person, clinicians will want evidence of those decisions: did it reach the target? did it activate? did it overactivate? The body is opaque; it hides its micro-events. Here, the ingenuity of synthetic biology shows up in the choice of reporters. Light-producing genes are a favourite in the lab, but human tissue is not a transparent petri dish. Newer approaches try to create signals that can be detected through the body. Acoustic reporter genes, for example, can produce gas-filled protein structures that scatter ultrasound, making microbial activity visible with a tool already present in many hospitals. The dream is not to turn patients into glowing lanterns, but to allow an ordinary imaging session to show whether a microbial therapeutic is alive, where it is, and whether its circuit has flipped from idle to active.

Delivery, too, resists simplistic solutions. A bacterium engineered to perform a function might still be destroyed by stomach acid, bile salts, and the immune system before it ever reaches its intended niche. In response, researchers have begun to treat the microbe almost like a spacecraft: it needs a heat shield for entry. Encapsulation materials and coatings can help a bacterium survive the passage through the upper gastrointestinal tract. More sophisticated still are designs in which the microbe's own surface properties change over time. One strategy demonstrated in mice uses inducible capsule expression: the bacterium temporarily cloaks itself to reduce immune detection during the vulnerable early phase, then sheds the cloak later to allow clearance. The capsule becomes part of the program. This is an important conceptual shift. The program is not only what the bacterium *does* at the target, but how it *travels* and how it *leaves*.

All these examples—PKU metabolism, inflammation memory, urine readouts, ultrasound visibility, programmable encapsulation—have the same underlying structure: *sense* $\rightarrow$ *compute/control* $\rightarrow$ *act*. Yet the hard part is not building a circuit that works once; it is building one that works reliably in the messy, evolving, deeply personal environment of a body.

The body differs between individuals in ways that matter to microbes. Diet alters bile acids and gut pH. Antibiotics and infections reshape microbial communities. Genetics and immune history influence which bacterial strains can persist. A living therapeutic that works in one person might be cleared too quickly in another, or persist too long. This variability pushes designers toward programs that are robust to context and toward dosing strategies that do not rely on indefinite colonization. Sometimes the safest design is not a permanent resident but a visitor: a strain that performs its task over hours to days and then disappears unless redosed.

That brings the discussion back to safety, not as a barrier but as the condition of possibility. The public imagination often treats engineered organisms as either miraculous or monstrous. Real engineered microbes are more prosaic and therefore more challenging: they are products, meant to be manufactured consistently, stored, administered, monitored, and recalled when needed. Their safety is not only about whether they are pathogenic. It is about whether they can share genes with other microbes, whether they can evolve away from their intended behaviour, whether they can persist in the environment after excretion, and whether their sensing logic can misfire in unexpected physiological states. Even the best containment strategies can behave differently outside the lab. A kill switch that appears reliable in one nutrient condition can fail more often in another. That is not a reason to abandon the idea; it is a reason to test containment in the contexts that matter, and to design with pessimism about what evolution and ecology will try.

The most persuasive vision of the programmable pill is therefore modest in tone but radical in implication. It does not promise a single microbe that cures all disease. It promises a new therapeutic medium: living cells that can be instructed, constrained, and audited. A decade ago, the notion that a bacterium could live in the gut long enough to act as a diagnostic device would have sounded like a stunt. Now the question is how to make such behaviour dependable enough for everyday medicine.

If that happens, the relationship between patient and drug changes. The patient no longer takes a substance that passively circulates. The patient hosts an agent that responds to local chemistry, one that can be designed to stay quiet until a very particular combination of cues appears. The most humane outcome of this technology would be a medicine that is both more targeted and less intrusive: a treatment that does not ask the whole body to pay the price for a problem confined to one tissue, a therapy that can be switched off as deliberately as it is switched on, and a diagnostic that can remember what the patient's symptoms cannot easily describe.

6.2 Sentinels of the Anthropocene

A canary in a coal mine was not a metaphor at first. It was a small, warm-blooded instrument carried underground because it breathed faster than a miner and succumbed sooner to carbon monoxide. When the bird's song faltered, men with lamps and pickaxes understood something invisible had changed. The canary offered no chemistry lesson and no map of the hazard. It did something simpler and more valuable: it turned a hidden threat into a perceptible event.

The Anthropocene—the proposed name for an age in which human activity leaves planet-scale signatures in air, water, and soil—is thick with hazards that behave like carbon monoxide. Many are colourless and odourless. Some are present at parts-per-billion and still matter. Others appear only briefly, in pulses: a burst of pesticide runoff after a storm, a plume of heavy metals after a pipe corroded through, a bloom of cyanobacteria after an unusually hot week. Electronic sensors can be exquisitely precise, but they are also expensive, power-hungry, and selective about what they can detect. A sensor is usually built to detect one thing well. The environment, meanwhile, is not polite enough to send one thing at a time.

Living organisms are already the planet's original sensor network. A lichen records air quality in its tissues. A mussel accumulates toxins in a way a chemist can later read. A microbiome shifts when a pollutant arrives, even if no one is watching. Synthetic biology takes that natural sensitivity and tries to make it legible on demand. Instead of waiting for an ecosystem to sicken and then inferring the cause, engineered organisms can be designed to report the presence of a threat and, in some cases, begin dismantling it.

The logic that makes a programmable pill plausible also makes a programmable sentinel plausible. Sensing is not a special feature that must be bolted on; it is what cells do for a living. A bacterium survives by constantly asking: is there oxygen here, is there sugar, is this temperature safe, are there toxins? The trick is to rewire those questions so that a human receives the answer, and to do so without creating a new ecological problem.

A useful starting point is heavy metals, because they are both ancient and modern. Arsenic, mercury, and lead have been moving through Earth's crust since long before humans mined them. What has changed is their concentration and their routes into the human body. Arsenic contamination of groundwater, for example, has harmed millions of people in South Asia, and smaller-scale problems exist worldwide. Standard testing is accurate, but it requires supply chains, trained personnel, and repeat sampling. It often arrives too late, after people have already been drinking the water.

Bacteria, however, have been dealing with arsenic for billions of years. Many species carry arsenic-responsive genetic systems that turn on detoxification genes when arsenic is present. Engineers can hijack that switch. If a promoter (a DNA control region) is activated by arsenic, it can be placed in front of a reporter gene—something that produces colour, fluorescence, or an electrical signal. The bacterium becomes a biological transducer: arsenic in, signal out. The dream of the field-deployable biosensor is not a laboratory curiosity glowing under a microscope, but a robust device a community can use.

Recent work has pushed in that direction by integrating engineered microbes with electronics. One striking example used *Bacillus megaterium*, a bacterium that can form spores, paired with CMOS (the same class of semiconductor technology used in many integrated circuits). Spores solve a practical problem that often kills good ideas at the border between lab and world: storage. A living sensor is useless if it must be refrigerated or fed carefully like an aquarium fish. Spores can be dried, transported, and revived when needed. Electronics solve the other half: they make outputs readable and standardizable, turning a biological response into a number or a simple pass/fail indicator. The hybrid device starts to resemble an appliance rather than an experiment.

This marriage of living and electronic systems changes the aesthetics of environmental monitoring. Traditional instruments are sealed, sterile, and inert. A biosensor can be porous and reactive, closer in spirit to soil itself. That intimacy with the environment is precisely what gives it power, but it also makes interpretation hard. A bacterium does not merely detect arsenic; it detects arsenic *as experienced by a cell.* Chemistry students learn to distinguish "total concentration" from "bioavailable concentration," because metals can be locked up in minerals or bound to organic matter. A living sensor naturally reports the biologically relevant fraction. That can be a feature, but it can also confuse: a water sample might have a high total arsenic number in a laboratory assay but a lower biosensor response if much of it is not accessible to organisms. The device does not lie; it answers a slightly different question. Learning to ask the right question is part of the technology.

The same subtlety appears in efforts to detect explosive residues and landmines. There is a long history of training animals for scent detection, from dogs in airports to rats used for mine detection in some countries. The appeal is similar: a biological nose is sensitive, cheap to replicate, and able to work amid complex background smells. Synthetic biology proposes a different kind of nose: bacteria engineered to respond to chemicals associated with explosives, producing a detectable signal. In principle, such bacteria could be deployed over a wide area, forming a distributed sensor field. In practice, the obstacles are formidable. Soil is harsh. The signal must be readable at a distance. The engineered organism must not spread. Yet even the imperfect attempts have value because they force clarity about what "detection" means: is it an alarm in real time, or a map built from samples collected over days? The answers shape design.

If biosensing is the canary's song, bioremediation is the canary learning to metabolize carbon monoxide.

Oil spills are the historical anchor for this idea. After the Exxon

Valdez disaster in 1989, the public saw, in television images, how a pollutant can rearrange an ecosystem. Less visible was the biological response: microbial communities shifted, and oil-degrading bacteria bloomed where hydrocarbons were abundant. Microbes had been eating hydrocarbons long before humans drilled wells; petroleum is a natural substance that seeps from the Earth. What a spill changes is scale and location. The hope of bioremediation is to accelerate a process that exists anyway, by giving microbes better enzymes, better access to the pollutant, or better survival in damaged habitats.

Plastic, however, is a different beast. Polyethylene terephthalate (PET), the polymer used in many bottles and fabrics, is designed to resist decay. Sunlight and mechanical abrasion can fragment it, but "fragment" is not "vanish." Microplastics persist and travel: in rivers, oceans, soils, and even the air. The problem is not only visual litter but chemistry and biology. Microplastics can carry additives, concentrate pollutants on their surfaces, and enter food webs. They also create new microbial habitats, sometimes called "plastispheres," with their own ecological dynamics.

A turning point in the public story arrived when researchers reported an enzyme, later dubbed PETase, that could break down PET into smaller molecules. The discovery suggested that even recalcitrant polymers might have biological vulnerabilities. Subsequent work explored variants of PETase and partner enzymes such as MHETase, aiming to improve speed, stability, and performance under industrial conditions. The "under industrial conditions" clause is crucial. It is easy to demonstrate activity in a purified assay in a warm, buffered tube. Wastewater sludge is not a warm, buffered tube. It contains detergents, oils, metals, variable pH, competing microbes, and mechanical shear. If biological recycling is to become more than a headline, enzymes must work where plastic actually accumulates.

Wastewater treatment plants have therefore become a strategic target, almost a planetary choke point. Modern cities funnel their microplastic output into these systems: fibres from laundry, fragments from packaging, particles from tire wear carried by runoff. Treatment plants already concentrate solids and manage microbial communities to reduce organic waste. They are controlled environments compared with rivers and oceans, and they are places where intervention can be monitored. If engineered enzymes or organisms can capture and degrade microplastics there, the benefits scale quickly.

There are different philosophies for doing this. One is to use purified enzymes, immobilized on surfaces or added as reagents. This avoids releasing living engineered organisms but can be expensive and requires replenishment. Another is to use whole-cell catalysts: engineered microbes that express plastic-degrading enzymes on their surfaces or secrete them, turning the cell into a self-replicating enzyme factory. Whole cells can, in principle, regenerate the catalyst and adapt to conditions. They also raise the sharpest safety questions.

Here the earlier lessons about "programmable" behaviour become environmental ethics. A bacterium in a gut is contained by anatomy and medical supervision. A bacterium in a river is contained only by design choices and enforcement of deployment protocols. The most thoughtful proposals assume that any engineered organism released into the environment should carry multiple layers of containment. Auxotrophy becomes one layer: the organism depends on a nutrient not found in nature. Kill switches become another: circuits that trigger death if the organism leaves a defined chemical context or after a certain number of divisions. Physical containment—encapsulation in beads, anchoring to filters, immobilization on membranes—adds an additional boundary, aiming to keep living agents within a device rather than free in a habitat.

A sobering finding in recent years is that containment performance can be strongly environment-dependent. A kill switch tested in standard laboratory media may show impressively low escape rates, suggesting that mutants that disable the switch are rare. But environmental conditions can change the arithmetic. Nutrient limitation, pH shifts, temperature swings, or chemical speciation can alter stress responses, mutation rates, and gene expression patterns. A trigger molecule might be present in the lab but absent in the field. A promoter might leak more in one context than another, creating selective pressure for mutations that silence it. In other words, the environment is not only where containment must function; it is a force that shapes whether containment remains intact.

This has an uncomfortable implication: biosafety is not a property of an organism alone, but of an organism in a particular setting. A responsibly engineered microbe is not "safe" in the abstract. It is safe, or unsafe, in a wastewater reactor with known chemistry, flow rates, and monitoring, and it may behave differently in a pond after a storm. That pushes the field away from grand gestures—"release plastic-eating bacteria into the ocean"—and toward constrained deployments: reactors, filters, and contained treatment stages where conditions can be designed as part of the safety system.

The dual-use problem shadows this work. The same techniques used to build microbes that detect toxins can, in principle, be used to build microbes that evade detection. The same knowledge that improves persistence of engineered strains can be misused to make harmful strains harder to eliminate. This does not mean that environmental synthetic biology is inherently reckless. It means that governance must be treated as an engineering parameter, not an afterthought. Controls over DNA synthesis, careful publication norms around potentially dangerous details, and institutional oversight are part of the scaffold that allows beneficial uses to proceed without naivety.

It is worth pausing on why anyone would accept the risk at all. The answer is that the risks of inaction are not hypothetical either. Plastics accumulate. Heavy metals do not degrade. Some pollutants move across borders without passports. And the costs of monitoring at high spatial resolution are enormous. A single electronic sensor buoy can be thousands of dollars and require maintenance. A living sensor can be grown from a small starter culture. If engineered responsibly, organisms offer a kind of scalability that electronics struggle to match: they copy themselves.

That self-copying, however, is a bargain with a trick clause. A distributed network of living sensors is tempting precisely because it can spread. Yet spreading is what makes ecological disruption possible. The safest designs therefore often constrain replication, even if it limits performance. A sentinel organism might be engineered to survive only briefly, long enough to signal, and then die. Or it might be locked inside a cartridge through which water flows, so that the "organism" is part of an instrument rather than an inhabitant of a habitat. The phrase "deploying engineered organisms" can conjure images of releasing swarms into the wild. Much of the most realistic work looks more like plumbing and industrial design: fitting biology into the interfaces where waste streams already pass.

The human story of these sentinels is not only about cleverness but about humility. Environmental problems are notorious for unintended consequences. Pesticides that saved crops also harmed birds. Refrigerants that cooled homes also thinned ozone. Any intervention that acts at scale can do harm if it is deployed with the confidence of a single-discipline solution. Synthetic organisms, in the best case, encourage a more integrated way of thinking because they force collaboration between molecular biologists, ecologists, engineers, regulators, and communities who will live with the results.

That collaboration also reshapes what counts as "success." A biosensor that fluoresces beautifully in a laboratory may fail if the fluorescence is masked by turbidity, algae, or sunlight. A plastic-degrading enzyme that is fast in a buffer may be irrelevant if it cannot access the polymer surface in real sludge. A kill switch that seems rigorous may be undermined if the trigger chemistry behaves differently in brackish water. These are not embarrassing footnotes. They are the core constraints that turn an interesting organism into an effective tool.

The Anthropocene is sometimes described as an era of planetary management, but management implies control, and control is often an illusion. A more honest ambition is stewardship: detecting harm early, localizing interventions, and designing tools that fail safely. Engineered organisms can be part of that toolkit if they are treated not as replacements for environmental responsibility but as instruments that widen the senses and extend the reach of remediation.

There is a quiet beauty in the idea that a microbe might patrol a stream for arsenic or sit in a wastewater tank dismantling the chemical stubbornness of plastic. It is a reversal of a familiar narrative in which microbes are enemies to be sterilized away. The same forms of life that once terrified cities through cholera and typhoid can also be domesticated into guardians, provided their autonomy is bounded and their behaviour is legible. In a world where so many threats are invisible until they become catastrophic, the most valuable machines may be the ones that can live in the mess, listen to chemistry, and answer with a signal we can finally hear.

6.3 The Architect of Immunity

The immune system is often described as an army, but that metaphor hides what makes it so strange and powerful. Armies follow orders from a central command. Immune cells do not. They are closer to a society of roaming specialists, each carrying a set of instincts encoded in genes and refined through experience. They negotiate with tissues, communicate by chemical whispers called *cytokines*, and make decisions locally. When they fail, the failures are not minor. Cancer is, in part, a failure to recognize a dangerous neighbour. Autoimmune disease is recognition without restraint. Allergies are the wrong kind of enthusiasm.

The most audacious idea in modern medicine is that these decisions can be rewritten. Not merely nudged by a drug or boosted by a vaccine, but reprogrammed by giving immune cells new rules for recognizing threats and new instructions for what to do when they find one. This is not a metaphorical "training" but literal genetic editing and synthetic circuitry. If the programmable pill placed engineered microbes into the body to sense and act, immune-cell engineering turns the body's own agents into living machines, built from a person's cells, tailored to a person's disease, and released back into the bloodstream with a mission.

It helps to start with what a T cell is. T cells are a class of white blood cell that patrols the body looking for signs of infection or malignancy. They do not typically recognize whole bacteria or whole tumour cells the way antibodies might. Instead, they inspect fragments. Almost every cell in the body displays tiny pieces of its internal proteins on its surface, held in molecular frames called *MHC* (major histocompatibility complex). A healthy cell shows the immune system a steady slideshow of "ordinary self." A virus-infected cell shows viral fragments. A cancer cell may show mutated fragments or abnormal levels of certain proteins. A T cell's natural receptor, the *T cell receptor* (TCR), recognizes these fragments in a highly specific way.

This system is elegant but not always convenient for therapy. Many tumours do not display strong, distinctive fragments. Some reduce MHC display to hide. Others thrive in tissues where immune responses are suppressed. The immune system is not blind to cancer, but it is often cautious, hesitant, or exhausted. The therapeutic question becomes: can a T cell be given a different way to see?

A historical anecdote captures the mood shift. In the late nineteenth and early twentieth centuries, William Coley, a surgeon in New York, observed that some cancer patients who developed severe bacterial infections occasionally experienced tumour regressions. He pursued this grim clue, deliberately injecting mixtures of killed bacteria—"Coley's toxins"—in an attempt to provoke an immune assault on tumours. The approach was inconsistent and dangerous, and it eventually fell out of favour as radiotherapy and chemotherapy matured. Yet the underlying intuition survived: the immune system can, under certain conditions, destroy cancer. The problem is control. Coley was trying to pull a lever with a sledgehammer. Modern immunotherapy tries to build a lever with a hinge.

The hinge is the receptor.

A *CAR-T* cell is a T cell engineered to express a chimeric antigen receptor (CAR). "Chimeric" means stitched together from parts that do not normally meet. On the outside, the receptor uses an antibody-like binding domain that can recognize a molecule on the surface of a tumour cell directly, without relying on MHC presentation. On the inside, it carries the signalling machinery that activates the T cell when binding occurs. In effect, the CAR turns the T cell into a hunter that recognizes a chosen target and, upon contact, shifts into an activated state: proliferating, releasing toxic molecules, and recruiting other immune activity.

The success of CAR-T therapy in certain blood cancers is one of the most striking achievements of personalized medicine. In diseases such as some B-cell leukemias and lymphomas, CAR-T cells targeting CD19 (a marker found on many B cells) have produced remissions in patients who had exhausted other options. The results can be life-changing, and they are also a reminder that changing an immune cell's decision-making can be as dangerous as it is powerful. A T cell that suddenly regards a large population of cells as enemies can unleash a storm.

That storm has a name: *cytokine release syndrome*. When CAR-T cells activate strongly, they and other immune cells can pour cytokines into circulation, producing fever, low blood pressure, and organ stress. Another risk is neurotoxicity, which remains an area of active investigation. These are not rare footnotes; they are part of the price of making the immune system decisive. The clinical craft has become, in part, the craft of managing an engineered living drug that can multiply inside the patient.

The workflow of CAR-T therapy also reveals why immune engineering feels like architecture. Unlike swallowing a pill, CAR-T therapy begins with extraction. Blood is drawn, and T cells are separated. Those cells are then genetically modified in the laboratory to insert the CAR gene, often using viral vectors that are good at delivering genetic cargo. The engineered cells are expanded—grown into a large population—and then reinfused into the patient, who may receive chemotherapy beforehand to make space for the incoming cells. The therapy is a loop from body to lab and back, and that loop introduces time, cost, and manufacturing complexity. Yet it also introduces a kind of accountability: the cells are handled, counted, tested, and monitored. They are an engineered product made from a person's tissue.

Even so, CAR-T is only the beginning of immune programmability. The deeper ambition is to make immune cells not merely targeted, but *conditional*—to give them the biological equivalent of logic gates so that they act only in the right context, with the right combination of cues, at the right intensity.

Cancer provides an obvious reason for this. Many of the surface molecules that distinguish tumours are not unique to tumours. They may appear at lower levels on healthy tissues. This creates the central safety problem of targeted immune therapies: *on-target, off-tumour* toxicity. The receptor finds what it was designed to find, but the target exists in a place you did not want attacked. The immune system's natural tolerance mechanisms evolved to avoid this; engineering must recreate it in a different form.

One approach is to demand multiple confirmations before a T cell unleashes its full program. This is where synthetic biology meets immunology most clearly. A cell can be built so that recognition of antigen A does not immediately trigger killing, but instead triggers expression of a second receptor or a second activating signal. Only when antigen B is also present does the full cytotoxic response occur. The underlying logic is similar to microbial *AND* gates, but the stakes are higher and the biology is different: the T cell's activation pathways are complex, and the consequences of an error are immediate tissue damage rather than mis-timed enzyme release.

Synthetic receptors known as *synNotch* have become a particularly elegant tool for this kind of control. The basic idea is to separate recognition from response more cleanly than natural receptors allow. When a synNotch receptor binds its target, it can release a transcription factor into the cell—a protein that goes to the nucleus and turns on chosen genes. That means binding can be wired to almost any genetic program: expression of a CAR, secretion of a cytokine, production of a checkpoint inhibitor, or activation of a suicide switch. The immune cell becomes a platform in which "if you see this, then do that" can be written in DNA, not just inferred from biology's default wiring.

This conditional design also addresses one of cancer's most infuriating tricks: heterogeneity. A tumour is not a uniform mass of identical cells. It is a shifting population, with subclones that vary in their surface markers and their susceptibility to attack. If a therapy targets a single marker, the tumour can sometimes escape by selecting for cells that do not display it. Immune circuits that can adapt—for example, by triggering secretion of factors that recruit other immune cells, or by enabling recognition of multiple targets—are attempts to keep pace with the tumour's evolution. The tumour is an evolving adversary; the therapy cannot be static.

At this point, gene editing enters the story. To "program" an immune cell is not only to add a new receptor, but sometimes to remove or rewrite parts of the cell's native machinery. The most famous editing tool, CRISPR, acts like a molecular word processor: a guide RNA directs a cutting enzyme (often Cas9) to a chosen DNA sequence, and the cell's own repair processes then mend the cut, creating a change. In practice, this can be used to knock out genes that limit T cell activity, to reduce the risk of the engineered cells attacking the patient's normal tissues, or to make the cells more persistent in hostile tumour environments.

First-in-human trials of CRISPR-edited T cells have provided a different kind of proof than the spectacular remissions of CAR-T. They demonstrate feasibility: that multiple genetic edits can be made, that edited cells can persist and be tracked, and that clinicians can monitor for safety signals such as unintended chromosomal rearrangements. This is a quieter kind of progress, the sort that does not always make headlines but determines whether the field matures. Editing immune cells is not merely an act of creation; it is an act of quality control. Each edit is a potential benefit and a potential hazard.

The hazards are not abstract. DNA cutting carries the risk of off-target edits, where sequences similar to the target are cut accidentally. Even on-target cutting can sometimes lead to unwanted rearrangements if repair goes awry. Moreover, immune cells are not passive vehicles. They have their own survival drives, and edited cells that gain a growth advantage could expand in ways not intended. This is why modern cell therapies increasingly include safety features: engineered "off switches" that allow clinicians to eliminate the cells if toxicity becomes severe. Some switches are based on a gene that makes the cells vulnerable to a particular drug; administering the drug triggers cell death. It is an emergency brake, and its existence is a reminder that engineered immunity is powerful enough to need one.

Another lesson comes from the body's own checkpoints. T cells have inhibitory receptors such as PD-1 that act as brakes, preventing excessive damage during infection and maintaining tolerance. Many tumours exploit these brakes by displaying ligands that keep T cells exhausted. Checkpoint inhibitor drugs, which block these inhibitory interactions, have transformed treatment for some cancers. But systemic checkpoint inhibition can also unleash autoimmune-like side effects, because the brakes are removed broadly. Engineered T cells can, in principle, localize such effects. A cell might be programmed to secrete a checkpoint-blocking antibody only when it has detected tumour signals, concentrating the brake-release at the site of disease rather than across the entire body. This is an example of immune-cell programmability as a strategy for *local pharmacology*: not changing what a drug does, but changing where and when it is made.

The emotional appeal of these therapies is obvious. They feel like justice. Instead of poisoning rapidly dividing cells indiscriminately, one recruits the body's own surveillance system and gives it clarity. Patients and clinicians often describe CAR-T cells as "living drugs" because, unlike conventional drugs, they can expand in the body, persist, and sometimes provide long-term control. The phrase is accurate, but it can mislead. A living drug is not merely alive; it is adaptive. It can change, and it can surprise. Managing that adaptiveness is the central challenge of the architect of immunity.

Architecture also implies constraints. Not every disease can be solved by unleashing cytotoxicity. In autoimmunity, for example, the immune system attacks healthy tissues. Here, the desirable engineered cell might be a regulator rather than an assassin: a T cell designed to calm inflammation, promote tolerance, or repair tissue. Some researchers are exploring engineered regulatory T cells (Tregs) that could home to specific tissues and suppress pathological immune activity. The same logic gates that restrict killing could restrict suppression, ensuring that immune dampening occurs only where it is needed. This is a subtler and, in some ways, more difficult form of immune engineering, because the goal is not to destroy a target but to restore balance without inviting infection or cancer.

The broader meaning of immune-cell programmability is that it turns therapy into a kind of software problem. Not because bodies are computers, but because complex behaviour can be specified in modular rules. Recognition modules, signalling modules, transcriptional modules, payload modules, safety modules. Once those modules are reliable, new therapies can be assembled by recombination rather than reinvention. This is the same industrial logic that transformed electronics: standardized components enabled rapid innovation. Biology is not as neat. Its components interact in context-dependent ways, and the immune system is a particularly context-dependent environment. Yet the direction of travel is unmistakable. Medicine is learning to write instructions into cells.

A person receiving engineered immune cells is, in a literal sense, receiving a modified part of themselves. That raises questions that are both practical and philosophical. Practically, who owns the therapy: the patient whose cells are used, the company that engineered them, the hospital that manufactured them? Philosophically, what does it mean to treat disease by altering the behaviour of your own living agents? The immune system is intimately tied to identity. It distinguishes self from non-self, remembers past infections, and shapes how the body interacts with its microbial companions. Rewriting it is not like replacing a broken bone. It is closer to changing the terms of an internal social contract.

The deepest promise of this work is also the most unsettling: moving from treating symptoms to rewriting causes. In some inherited diseases, the immune system is not the problem, but immune cells can be used as vehicles for gene editing or as targets for correction. In cancer, engineered T cells can sometimes achieve what chemotherapy cannot. And beyond cancer, immune programmability offers a blueprint for responding to threats that evolve quickly: viruses, perhaps, and even the shifting pathologies of aging tissues.

For now, the clearest achievements remain in the careful, high-intensity setting of specialized clinics, where patients can be monitored closely and interventions can be adjusted in real time. That is fitting. When one starts to build with the immune system, one is building with fire. But fire is not only destructive. It can be contained, directed, and used to shape tools that were impossible in cold darkness. The most remarkable aspect of immune-cell engineering is not any single remission or any single receptor design. It is the emerging sense that immunity is not merely something we have, but something we can, with enough care, learn to compose.

Chapter 7

Pandora's Box and the Kill Switch

Synthetic biology promises to rewrite the rules of life, yet the specter of escaped organisms and weaponized pathogens looms large. This chapter investigates the molecular safeguards designed to keep artificial life contained-from genetic firewalls to metabolic addictions-and confronts the precarious balance between open scientific innovation and global biosecurity.

7.1 The Fortress and the Fail-Safe

Laboratory safety historically begins with walls: concrete rooms, glass cabinets, HEPA filters, negative-pressure suites and access badges. Those fortifications are indispensable. They are also, in a literal sense, guardians against what we can see. A fume hood catches a splash; a biosafety cabinet intercepts an aerosol. But living systems are microscopic, mutable and ingenious at slipping past the obvious. A spore survives heat. A plasmid slides between bacteria. A single point mutation, occurring by chance, can silence an engineered control. When the opponent is a replicating molecule that can change its own instructions, a fence—no matter how high—can be scaled, tunneled under, or simply ignored when human error opens the gate.

Containment for microbes has always therefore been two things at once: physical and procedural. But physical barriers address exposure, not the fundamental property that makes biology both useful and dangerous—life replicates. If an engineered organism escapes a lab and finds conditions in which it can reproduce, it may adapt. The design response to that reality has been to fold safety into the organisms themselves: to rewrite some of their most basic dependencies so they cannot live outside of carefully provisioned conditions. A fortress remains useful, but the true fail-safe is a genetic commitment to dependence.

One straightforward idea borrowed from classical genetics is *auxotrophy*. In nature, an auxotroph is a microbe that cannot make an essential nutrient and must import it from the environment. A strain that cannot synthesize the amino acid tryptophan, for example, will die if tryptophan is absent. Synthetic biologists turned that old lesson into an engineered addiction. Instead of relying on naturally available nutrients, they reprogram essential cellular machinery to require a molecule that exists only inside the laboratory. The organism, in effect, becomes a controlled consumer: it flourishes when supplied with a supplied chemical, and without that chemical it simply cannot survive.

Making such metabolic dependencies robust is not a trivial matter. Early auxotrophies were useful for laboratory selection but could be rescued by a variety of ecological tricks. Bacteria exchange metabolites; a needy strain co-habiting with a generous wild microbe can be fed. Mutations can restore pathways or reduce dependency. Horizontal gene transfer—bacteria picking up genetic material from their neighbours—can supply the missing capability. To confront those escape routes, engineers layered strategies.

One influential approach was to make the dependency molecularly exotic. Instead of asking the cell to need a common vitamin or amino acid, scientists rewired essential proteins so that they physically incorporate a non-natural amino acid—one that does not exist in nature and is manufactured only for the laboratory. If an essential enzyme will function only when one of its positions is occupied by this synthetic amino acid, the cell becomes literally unable to assemble that enzyme without the laboratory supply. Work from several groups in the mid-2010s demonstrated that recoding multiple essential genes to require such an *unnatural* residue could create very tight containment: the organism can survive and grow in a supplemented fermenter, but it dies when released into ordinary environments that lack the synthetic building block. Rovner and colleagues (2015), and related teams around the same years, showed how redesigning proteins to depend on these supplied amino acids made escape highly unlikely unless multiple coordinated changes occurred.

These systems are the modern equivalent of a laboratory key: remove the key and the organism stops. But keys can be duplicated. Designers therefore borrowed ideas from security engineering: combination locks rather than single keys, redundancy rather than single points of failure, and active kill switches that do not merely withhold sustenance but actively destroy cells that stray. Two well-known circuit architectures—dubbed *Deadman* and *Passcode*—illustrate how genetic logic can tie survival to specific environmental inputs. Deadman works like a countdown: an engineered genetic circuit produces an essential factor only while a supplied chemical is present; if the chemical is withdrawn, a built-in timer runs out and the cell self-destructs unless the signal is renewed. Passcode circuits layer multiple inputs so that survival requires the correct combination—an engineered microbe might need two or three specific small molecules presented in a defined temporal pattern to remain alive. These designs reduce the probability that a single mutation or accidental exposure will restore fitness outside the intended context. Stirling and colleagues (2017) described such circuits, and subsequent work explored ways to harden them against simple mutational inactivation.

Even so, evolution is patient. Toxin-antitoxin modules and single-component kill switches commonly select for mutants that delete, silence or otherwise evade the circuitry. Engineering teams learned to complicate the evolutionary pathway for escape: place essential functions under the control of multiple, orthogonal mechanisms; recode several independent genes; and design circuits whose failure is itself deleterious to the host. Another lesson was ecological: a population that escapes but is out-competed by wild-types will not flourish. Combining intrinsic molecular constraints with ecological pressure—the engineered strain loses a competition it cannot win without the supplied support—provides another layer of defense.

Beyond metabolic addiction and circuit-based self-destruction, researchers pursued the idea of making the language of life itself foreign. The genetic code—the rules that translate DNA triplets into amino acids—was long thought to be universal. Altering that universality creates a *semantic* firewall: recode a host so that particular codons mean different amino acids, or remove codons entirely and reassign them, and foreign genes arriving by horizontal transfer are likely to be mistranslated and nonfunctional. Lajoie and collaborators (2016) carried out broad genomic recoding in bacteria and showed that such recoded organisms are less hospitable to viruses and certain mobile genetic elements. Later efforts pushed further, swapping codon meaning to actively mistranslate invading proteins, producing a form of built-in immunity to many natural parasitic elements.

Semantic containment is elegant, but it is not invulnerable. Mobile genetic elements sometimes bring their own translation tools—

tRNAs, for example—that can restore compatibility with the invading sequence. In practical terms, that means one cannot rely on a single firewall; recoding buys time and resilience but must be paired with other restrictions. The most effective designs are layered: metabolic addictions that create absolute dependencies; circuit safeguards that kill or disable cells that stray; and informational modifications that reduce the likelihood of gene flow and viral infection.

A concrete historical example makes these abstractions tangible. In the mid-2010s, teams sought to manufacture safer chassis organisms for industrial biotechnology—microbes intended to perform tasks outside the tightly controlled fermenter, such as bioremediation or agricultural assistance. Two lines of attack gained prominence. One recoded essential genes to require synthetic amino acids. Another built kill switches based on combinatorial inputs, so that a microbe in a field would require not just a nutrient but a sequence of signals available only under company control. When these ideas were first piloted, the community saw both promise and the sobering reality of escape routes: even well-designed auxotrophies could be rescued by co-occurring microbes, and simple kill switches could be lost through mutation. Those early prototypes therefore drove a change in ethos: rather than attempting a single silver-bullet containment, teams began to design for *defense in depth*—multiple, independent mechanisms that would need to fail simultaneously for an escape to succeed.

That change in approach also altered how engineers think about failure modes. Instead of considering the organism as a static device, they model it as an evolving population interacting with environments and other species. That perspective motivated experiments that ask, in effect, how many mutational steps are needed to restore growth without the laboratory supply, or how readily a missing function could be supplied by nearby microbes. The answers inform design choices: prefer dependencies on molecules unlikely to be produced by environmental microbes; recode multiple independent genes rather than a single locus; and design kill switches that target both the host genome and potential mobile elements that might carry copies of the engineered cassette.

Advances in gene editing have given new tools to enforce containment. CRISPR, the adaptive immune system repurposed for editing, can be flipped from a tool of modification to a mechanism of policing. A CRISPR-based guard can be programmed to recognize the engineered construct itself and cut it if certain conditions are met, ensuring that the genetic payload cannot be easily transferred to a wild host. Recent work has combined metabolic auxotrophy with CRISPR-based devices that selectively destroy bacteria regaining lost capabilities, closing a historically weak escape route—horizontal gene transfer. The message is clear: containment is increasingly an active process—cells not only fail to grow outside the lab but are targeted for elimination if they attempt to abandon their engineered constraints.

All the technical ingenuity, however, collides with practical limits. Some containment strategies are costly in growth rate or yield—an engineered industrial microbe that grows too slowly under production conditions is economically unattractive. Designers therefore balance safety and performance, sometimes sacrificing extreme stringency for viability. Trade-offs also arise when scaling from controlled lab conditions to messy real environments. The more an engineered strain needs human-provided inputs, the less useful it may be for an open-environment application. Conversely, loosening dependencies to increase field utility increases the burden on biosafety engineers to layer compensating safeguards.

Human factors reappear at every turn. A containment design that is flawless in concept can be undone by a labeling error, a forgotten reagent, or a misapplied protocol. Institutional oversight and culture matter as much as genetic circuits. The containment toolkit, therefore, is most effective when decentralized into multiple layers: physical barriers, trained people, audited procedures and genetically encoded fail-safes working together. The laboratory that relies on a single method—whether a glass wall or a single kill switch—invites catastrophe.

It is worth tracing one illustrative pathway through a realistic scenario. Imagine a company deploying an engineered bacterium in soil to break down an industrial pollutant. The strain carries a synthetic-auxotrophy circuit requiring a non-natural amino acid for critical enzyme function; it also harbors a Passcode-style circuit that only allows survival in the presence of a particular combination of small molecules supplied in a controlled formulation. Production occurs in a closed fermenter, where the supply is guaranteed. During transport, a containment breach dumps a small volume of culture into the environment. Alone, a handful of cells will face immediate lack of the synthetic amino acid and the Passcode inputs, and the default expectation is death. But if the environment contains organisms capable of producing a chemical that partially substitutes for the synthetic requirement, or if horizontal transfer occurs that supplies a missing function, a few survivors might arise. If the design included recoded genes that mistranslate foreign elements and a CRISPR guard that eliminates any cell acquiring the engineered cassette in an unintended host, the chance of a viable lineage establishing itself plummets. If, instead, the strain relied on a single auxotrophy and a single toxin-antitoxin switch, the evolutionary path for escape could be far shorter. The layered failure probabilities multiply: each independent safeguard reduces risk exponentially when they are truly independent.

The shift toward genetic solutions also changes the conversation about where responsibility lies. A glass-fronted containment facility is tangible and easily audited. Genetic containment, being invisible to the unaided eye, requires different verification: genomic audits, challenge experiments, long-term evolutionary assays and independent certification. These forms of evidence can reassure regulators and the public only if they are transparent and replicable. Engineers therefore prototype, measure, and report the evolutionary stability of their designs: how many generations were run, what stressors were tested, and whether any escape mutants were observed. The community's comfort with deploying living technologies increasingly depends on such demonstrations of robustness under realistic evolutionary pressures.

A final practical point concerns reversibility. Safe failure modes can be active—cells that are triggered to die—or passive—cells that simply fail to grow. Active elimination mechanisms are attractive when one wants on-demand removal, as when an engineered commensal must be cleared from an animal host after its task is complete. But active systems can also impose stronger selection for resistance. Passive systems—metabolic dependence—yield a quieter evolutionary landscape but are vulnerable to ecological rescue. The best designs blend both: a metabolic addiction that curtails growth outside the lab, coupled with an active trigger that eliminates rare survivors or prevents the transfer of engineered elements to other organisms.

After decades of trying different recipes, a practical design philosophy has crystallized: assume failure, design for failure modes, and stack independent layers that address distinct vulnerabilities. The fortress remains useful but is no longer sufficient. Glass and steel buy time and reduce the probability of exposure; genetic fail-safes reduce the probability of establishment; institutional practices reduce the probability that an exposure becomes a crisis. When those layers are thoughtfully combined, the overall system is far more resilient than any single measure could be.

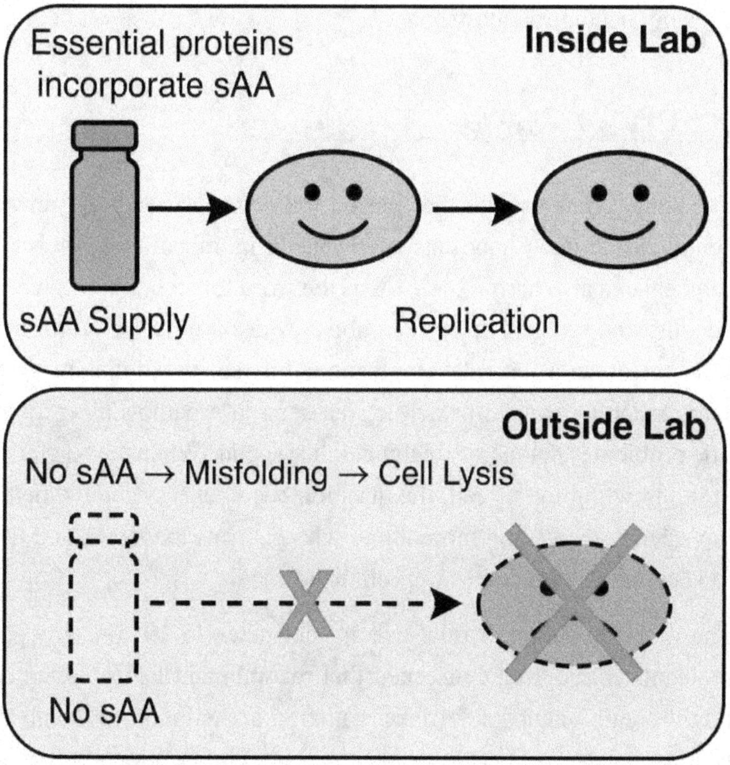

Genetic fail-safes are not an insurance policy guaranteeing zero risk. They are a change in posture: instead of assuming that the world can be perfectly walled off, designers accept that exposures will happen and insist that any exposed organism be extremely unlikely to survive, reproduce or share its engineered traits. The technical tactics—synthetic auxotrophy, recoding, circuit-based kill switches, CRISPR policing—are tools in that posture. What keeps them useful is continuous testing against realistic challenges: evolutionary selection, ecological interactions, and human fallibility. Engineering trust into living machines requires the same iterative humility and rigorous trial that made modern aviation safe: learn from near misses, model failure paths, and bake redundancy into systems so that a single flaw does not become a catastrophe.

7.2 The Double-Edged Helix

The same breakthrough that placed the power to read and rewrite genomes into many hands also sharpened a moral paradox: tools that can heal can also harm. *Dual-use* is the word for technologies whose benefits and dangers are inseparable. A desktop DNA synthesizer can accelerate vaccine development and personalized therapies; the same machine can, in principle, reassemble a pathogenic genome. The problem is not merely technical; it is social. When the barriers to building with biology fall, the question becomes less about whether the science works and more about who controls the levers and how to keep the bad outcomes improbably difficult.

The moral contours of dual-use are not new. In 1975, a group of molecular biologists, concerned that recombinant DNA techniques might create unknowable risks, gathered at Asilomar to weigh the promise of the field against public safety. The meeting set a precedent: scientists themselves would not only invent but also deliberate on the governance of their inventions. That reflex—mixing enthusiasm with restraint—has recurred whenever biology has leapt forward. But the context has changed. Recombinant DNA once required specialized facilities and a cadre of trained technicians. Today, CRISPR gene editing, high-throughput DNA synthesis and cloud-based design tools let small teams do in days what once took months in large institutions. The scale and speed of capability diffusion multiply both opportunity and hazard.

Concrete episodes bring the abstract into focus. In 2011 a pair of papers from two different groups—led by Ron Fouchier and Yoshihiro Kawaoka—reported that a highly pathogenic avian influenza virus (H5N1) could be made transmissible among mammals under laboratory conditions. The work aimed to understand what genetic changes might enable human-to-human spread; it also triggered an intense debate about whether such knowledge should be published at all. Some argued that detailed methods could serve as a blueprint for misuse; others insisted that withholding information would impede surveillance and preparedness. That controversy led to a de facto pause in certain lines of gain-of-function research and to tighter oversight—an early signal that dual-use dilemmas are as much about publication and policy as about benchwork.

Another wake-up moment came from gene synthesis. In 2017 a group reconstructed horsepox virus from chemically synthesized DNA fragments, producing a virus closely related to smallpox. The work had legitimate scientific rationale—understanding orthopoxvirus biology and vaccine design—but it also showed that a virus of historical consequence could be assembled from ordered DNA, raising fears about the potential to resurrect smallpox. The event exposed gaps in how DNA synthesis orders are screened and how international norms are enforced. Companies that manufacture sequences began to adopt voluntary screening protocols; researchers and funders grappled with whether and how to publish methods that might lower the bar for malicious actors.

Dual-use is therefore a technical problem wrapped in social choice. Technical responses exist, but they interact with incentives and values. One engineering path is to make biological tools intrinsically safer—built so that misuse is difficult or impractical. Two strands of such work are informative. One transforms the chemistry of living systems so that engineered organisms become incomprehensible or inoperable to natural life. The other focuses on controlling the digital and supply chains that make biological construction possible: screening sequence orders, monitoring synthesis requests, and instituting norms for responsible publication.

The chemical approach ranges from modest redesigns to radical reinvention. Genomic recoding—the wholesale reassignment or removal of codons—changes the meaning of DNA. If an organism is engineered to interpret a particular three-letter DNA sequence as a different amino acid, a gene borrowed from a natural microbe will not function correctly in the engineered host. Lajoie and colleagues pushed this idea forward with broad recoding experiments that removed specific codons and reassigned them, showing that such organisms can resist some viruses and prevent easy horizontal gene transfer. More recent strategies go further: reassigning codons to amino acids that are not among the standard twenty or swapping meaning so that invading proteins are mistranslated, rendering them nonfunctional. Such semantic firewalls reduce the risk that engineered DNA will function if it migrates into wild organisms, and they make the engineered host a poor vessel for propagating foreign genes.

Yet nature fights back. Mobile genetic elements—plasmids and phages—sometimes carry not only genes but also tools for translation, such as tRNAs, that can restore compatibility with foreign messages. Experiments have demonstrated that, given enough selective pressure, mobile elements can supply missing components and thereby undermine naive recoding defenses. That observation drives an increasingly sophisticated posture: make the code change deep and multifaceted, combine it with metabolic dependencies that cannot be easily supplied by the environment, and deliberately design redundancy so that restoring function would require many simultaneous, unlikely changes. Xenobiology, the most radical version of this strategy, envisions life built on fundamentally different biochemistry—alternative nucleic acids (XNA) and nonstandard amino acids—so foreign to natural systems that genetic exchange becomes implausible. Xenobiology is powerful in theory, but it is also technologically demanding and raises its own uncertainties about ecological interactions and unforeseen chemistry.

The second, and in many ways more immediate, axis of defense is governance of the digital and supply chains that underlie biology. In the early commercialization of DNA synthesis, orders could move from an email request to a delivered sequence without systematic checks. Awareness of the problem led to industry-led screening initiatives that compare requested sequences against databases of known harmful agents and flag problematic orders for human review. The pipeline—digital design, algorithmic screening, human oversight—adds friction and accountability. Yet it is imperfect: algorithms must balance false positives and false negatives; databases must be updated; and bad actors can attempt to evade screening by fragmenting orders or using novel sequences not present in the reference sets.

A curious tension arises here. On one hand, openness accelerates innovation: sharing sequences, tools and protocols lets scientists iterate quickly and detect errors. On the other hand, openness can reveal vulnerabilities. The H5N1 debate was not about the science alone but about the ethics of disclosure. Some have proposed tiered openness—broadly publishing conceptual results while holding back procedural details that materially lower barriers to risky experiments. Others favor full transparency accompanied by robust surveillance and strong social norms against misuse. Those are political choices as much as technical ones, and they reflect different attitudes toward trust in professional communities, the state and markets.

Democratization of biology has another, often overlooked, human dimension: community laboratories and hobbyists. Biohacker spaces—community biology labs—have made citizen science a reality, enabling enthusiasts to explore genetics, microbiology and fermentation. Most of this activity is benign and educational, but the spread of capability beyond traditional institutions complicates the enforcement of norms and rules. The response combines outreach (teaching biosafety and ethics), lightweight governance (codes of conduct and training), and technical measures (requiring orders for certain reagents to pass through established suppliers who screen requests). That balance matters because the aim is not to suppress curiosity but to channel it responsibly.

Legal and institutional frameworks have adapted unevenly. Existing export controls, biosafety regulations and funding restrictions capture some high-risk activities, but many innovations fall into gray zones. International differences in regulation create opportunities for jurisdiction shopping. The global character of science—cloud tools, distributed collaboration, cross-border synthesis orders—requires cooperative norms and interoperable technical systems. Soft law—community standards, corporate best practices, and trusted registries—has played an outsized role precisely because formal regulation struggles to keep pace with fast-moving technology.

That fast pace also changes the calculus for risk assessment. The old model—evaluate a project in isolation—struggles when a technique developed for therapeutic purposes can be repurposed in an entirely different context. Risk is not a static property of a method but a function of accessibility, intent and secondary applications. A useful analytic shift sees risk as an emergent property of the ecosystem: who can use technique X, how cheaply, and with what likely detection or mitigation? That perspective encourages measures that increase the cost and reduce the stealth of harmful use: vigilant synthesis screening, improved forensic capabilities to trace origins, and rapid diagnostic networks that limit the window during which a novel pathogen could spread undetected.

Forensic and attribution capabilities deserve attention because they change deterrence dynamics. If misuse is likely to be traced back to a source quickly, malicious actors lose plausible deniability and face higher perceived costs. Building robust genomic surveillance, consistent metadata standards for sequence provenance and better tools for distinguishing natural from engineered signatures all feed into a broader ecosystem that discourages misuse. Such capabilities are also politically sensitive; surveillance can infringe on privacy and civil liberties if not carefully constrained. Again, the tradeoffs are social choices.

Training and culture turn out to be as important as circuits and codes. Scientists are not merely technicians; they are the first line of situational judgment. Teaching a sense of responsibility—an understanding that particular experiments carry unique harms if misapplied—affects decisions about what to publish, who to collaborate with and how to design experiments. Codes of conduct, mentoring, institutional review boards with dual-use expertise, and reward structures that value safety-conscious behavior all tilt the field toward prudence without halting innovation.

The interplay of technology and policy is never purely defensive; it reshapes what is possible. If an engineering solution like genomic recoding becomes standard, it will change the kinds of organisms researchers choose to work with and may lower the need for some procedural controls. Conversely, if governance succeeds in making certain practices ethically or legally untenable, innovation will shift in directions that are safer by design—perhaps toward xenobiology or toward confined systems that never leave the factory floor. Such shifts are not predictable in detail, which is why continuous dialogue among scientists, ethicists, regulators and the public is essential.

Stories capture both the promise and the peril. A synthetic biology start-up might aim to detoxify an abandoned industrial site by releasing a microbe that metabolizes a stubborn pollutant. The same toolkit lets another actor imagine modifying a microbe to produce an unregulated toxic compound. Whether the end result is salubrious or malevolent depends on more than an individual engineer's intent; it depends on supply-chain controls (who can order genes?), surveillance (how quickly would an anomalous release be detected?), and culture (would colleagues raise concerns about a risky plan?). The technical solutions—recoding, metabolic locks, kill switches—can reduce the window of misuse, but they do not replace the need for responsible institutions and informed public debate.

Practical safeguards continue to evolve. Gene synthesis providers refine screening algorithms and adopt shared databases of sequences of concern. Funders and journals increasingly require risk assessments and compel authors to justify transparent publication. Community labs develop mentoring programs and supplier relationships that discourage dangerous experiments. At the same time, the frontier of safer-by-design research—semantic containment metrics, amino-acid-swapped codes, XNA systems—offers a longer-term path to making misuse materially more difficult. If life can be made deliberately exotic—chemically and informationally incompatible with natural systems—then the very act of engineering creates a barrier to malevolent repurposing.

The double-edged nature of biology means that no single fix will suffice. Technical ingenuity must be matched by careful policy, credible attribution, community norms and public engagement. That combination does not eliminate risk, but it reshapes the landscape so that the default of use is benefit, not harm. The point is not to freeze scientific curiosity, but to build architectures—technical, institutional and cultural—that make harm expensive, detectable and socially unacceptable. Only by attending as much to the social machinery of science as to the molecular machinery of life can it be possible to steer the helix toward outcomes that heal rather than wound.

7.3 Engineering Trust

Trust is an engineering problem as much as it is a moral one. When people worry about synthetic organisms, their fears are rarely about the pipette or the petri dish; they are about uncertainty, surprise and loss of control. Trust answers three simple questions for a skeptical public: Can the technology deliver benefits? Can those benefits be measured and distributed fairly? And who will be held accountable if something goes wrong? The technical promise of programmable life must therefore be matched by social designs that make those answers credible.

The history of biotechnology offers a pattern. When a new capability arrives, early adopters—scientists, entrepreneurs, regulators—tend to assume that technical safeguards and institutional norms will suffice. The public, however, judges by experience: broken promises, perceived secrecy, and visible harms linger. Consider the Golden Rice controversy. Developed to address vitamin A deficiency, Golden Rice was presented as a humanitarian solution; yet for decades it encountered prolonged public resistance, regulatory hurdles and polarized debate about who benefits and who bears the risks. The story was less about the rice itself than about failures of engagement: communities felt excluded from decisions, questions about corporate interests and intellectual property were left unanswered, and a narrative of outsiders imposing a technological fix hardened opposition. The lesson is blunt: technical merit alone does not produce social license.

Ethics and governance have matured in response. Institutional mechanisms—Institutional Biosafety Committees (IBCs), biosafety officers, and regulatory guidelines—create a baseline of accountability. Revisions to the NIH Guidelines in recent years have clarified expectations for certain kinds of work and emphasized institutional responsibility for risk assessment and containment. International bodies, including the World Health Organization, have developed frameworks that treat biorisk management as a shared responsibility across biosafety, biosecurity and public health, providing practical checklists and principles for implementation. These frameworks do not eliminate disagreements, but they shift the terrain: decisions become procedural and auditable rather than raw assertions of expertise.

Meaningful oversight, however, is not simply a checklist. It is a forum for dialogue. The phrase *upstream engagement* captures a practical change in posture: invite affected communities, ethicists, and stakeholders into deliberations before projects reach an advanced stage. That invitation matters for two reasons. First, it surfaces concerns that are not purely technical—questions about fairness, cultural values, or long-term stewardship that engineers may not anticipate. Second, it builds channels of reciprocity: when local communities see their values reflected in safeguards, they are more likely to partner rather than resist. Upstream engagement is not a marketing exercise; it is a reorientation of project design so that societal values shape technological parameters from the outset.

Transparency is part of trust but not identical to it. Posting protocols and data online is necessary but not sufficient. Transparency must be meaningful: it should enable independent verification and foster understanding among nonexperts. That requires translation—bringing technical details into a format that civic actors can evaluate—and verification—third-party audits or reproducible demonstrations of safety. Some organizations now publish independent safety assessments and evolutionary-stability data for engineered strains, offering the kind of empirical evidence that regulators and communities can interrogate. Those documents do the work of persuasion by replacing abstract assurances with concrete, measured outcomes.

A vital strand of engineering trust is traceability. If an engineered organism causes harm, the public rightly expects a timely answer: what went wrong and who is responsible? Designing organisms and processes with traceability in mind helps. That can mean molecular provenance—built-in genetic markers that allow forensic identification—or supply-chain provenance, where orders for critical reagents are logged and screened. The Office of Science and Technology Policy's 2024 framework, for example, pushed responsibility upstream by tying nucleic acid synthesis screening requirements to federal funding and procurement, signaling that traceable supply chains are part of the safety fabric. Such measures also bolster deterrence: the easier it is to trace a misuse, the less attractive misuse becomes.

Trust also demands attention to equity. The benefits of synthetic biology—disease prevention, environmental remediation, agricultural resilience—will be felt unevenly unless distributive questions are addressed. Who owns the organisms? Who profits from commercialized applications? Are communities hosting field trials given a voice and a share of benefits? Past controversies show that perceived or real inequities can corrupt public confidence faster than any technical failure. One approach is to embed benefit-sharing agreements and accessible licensing models into project proposals, making the economic and social commitments explicit from the start.

Ethics frameworks must also reckon with irreversibility. Some interventions—gene drives being the paradigmatic example—intentionally propagate through wild populations. The potential ecological consequences are vast and hard to predict. Where reversibility is low, the demand for broad societal deliberation becomes strongest. National Academies reports and international guidance emphasize layered governance for such cases: stronger institutional review, regional cooperation, independent risk assessment and public consultation. Designing trust for irreversible actions therefore means slowing down, widening participation and insisting on stronger evidentiary standards.

Practical trust-building also touches the incentives that shape behavior. Academia and industry reward novelty, publication and speed; safety and careful deliberation rarely receive comparable recognition. Realigning incentives—funding long-term safety research, rewarding reproducibility and valuing stewardship in tenure and investment decisions—tilts the system toward reliability. Policies that fund the testing of containment systems, or that require safety data before scaling a technology, create a market for prudence. A 2025 U.S. Executive Order attempted to sharpen these incentives by directing revisions to oversight frameworks and by emphasizing audits and transparency for high-risk work, signaling that funders and governments can change institutional priorities through procurement and regulatory levers.

Communication is both art and science. Technical explanations alone often backfire when audiences fear marginalization or misuse. Effective communicators tell human stories: how a technology will affect livelihoods, who stands to gain, and how risks are being managed. They acknowledge uncertainty rather than hide it—paradoxically increasing credibility by showing realism. When the He Jiankui affair became public in 2018, the visceral reaction was not merely about the technical misstep but about secrecy and an apparent disregard for ethical norms. The scandal demonstrated how rapidly trust can collapse when a line of professional and public confidence is breached. Rebuilding that trust requires openness, consequences for misconduct, and sustained engagement with communities and peers.

Citizen participation matters too. Community laboratories and biohacker spaces illustrate both the promise and challenge of democratized biology. They foster education and grassroots innovation, yet they also broaden the set of actors for whom norms and training must be applied. Successful community labs build cultures of safety by pairing low-cost training with codes of conduct, mentorship and partnerships with established institutions. That approach treats curiosity as an asset to be nurtured responsibly rather than a risk to be suppressed.

Regulation plays a balancing act. Heavy-handed rules can stifle innovation and push actors into opaque venues; insufficient oversight invites harm and erodes public confidence. The most effective regulatory regimes are adaptable, proportionate and designed to foster learning. They incorporate mechanisms for periodic review, require post-deployment monitoring and encourage data-sharing about near-misses. That last point is important: an environment that rewards reporting of near-misses without punitive overreaction produces a culture of continuous improvement.

Independent assessment bodies have emerged to provide credibility beyond the institutions that develop technologies. Third-party evaluations, certification schemes and public audits signal that safety claims have been scrutinized by impartial actors. Analogies exist in other sectors—airline safety and medical device regulation, for instance—where independent testing and public reporting underpin trust. For synthetic biology, developing robust standards for what constitutes an adequate safety dossier is an ongoing task. Shared benchmarks—measures of evolutionary stability, containment robustness and traceability—would enable more consistent comparison across projects.

Insurance and liability frameworks further knit technical design to societal expectations. If companies are insured against accidental releases only when they meet certain containment standards, market forces will favor safer designs. Liability regimes that clearly assign responsibility for harm also encourage organizations to invest in verification and redundancy. These economic instruments do not replace ethical duty, but they align incentives so that safety becomes good business.

Finally, there is a cultural dimension. Trustworthy practice combines humility with competence. Humility acknowledges that biological systems are complex and that surprises will occur. Competence manifests as rigorous testing, transparent reporting and a willingness to accept limits. Scientists who publicly articulate uncertainties and who pre-commit to monitoring and redress build gravitas. Communities notice when researchers prioritize listening and when institutions enforce consequences for shortcuts. Over time, repeated demonstrations of responsible behavior accumulate into reputational capital—the kind of long-lived trust that allows controversial but beneficial projects to proceed.

Engineering trust is therefore a design project with many moving parts: procedures that can be audited, communication that is candid and accessible, institutions that distribute responsibility and oversight, and technologies that are safer by design. When those pieces work together, people experience not just explanation but assurance: a sense that benefits will be realized, harms will be managed, and that they have a voice in decisions that touch their lives.

Chapter 8

Life, Version 2.0

As we master the tools of biological creation, the definition of life itself begins to blur. This final frontier moves beyond merely tweaking existing organisms to designing biological systems for the harsh vacuum of space and rewriting the fundamental genetic alphabet. We stand on the precipice of a new era where the distinction between the born and the made-and even the terrestrial and the alien-finally dissolves.

8.1 The Interplanetary Petri Dish

Long-duration missions turn a spaceship into a small, fragile economy. Every kilogram shipped from Earth costs thousands of dollars and a finite slot on a rocket; every resupply mission arrives months later, if at all. Carrying enough food, oxygen, medicines and spare parts for a multi-year stay on Mars is simply impractical. The only scalable alternative turns an obvious biological truth into an operational principle: living things are factories that can eat waste and make value. The ambition is not to replace engineers with microbes, but to design organisms that become reliable, compact, self-repairing machines—onboard factories that convert human waste, Martian CO_2 and sunlight into oxygen, nutrients, medicines and materials in a continuous loop.

Imagine the habitat as a miniature ecosystem whose primary job is faithful bookkeeping: track inputs and outputs, recycle whenever possible, and supply whatever the crew needs with minimal intervention. Photosynthetic microbes—carefully engineered cyanobacteria or algae—capture sunlight and carbon dioxide to generate oxygen and fixed carbon. Heterotrophic microbes convert human and plant waste into simple substrates that production strains can turn into protein, vitamins and specialty molecules. Electrochemical interfaces let microbes produce fuels or feedstock chemicals by coupling metabolic pathways to electrodes. Modular bioreactors, microfluidic control systems and automated analytical instruments create a feedback loop: sensors measure nutrient levels and contaminants, controllers adjust feeds and light, and algorithms select which engineered strains to grow, when, and at what density. The result is a closed-loop life-support system that treats biology as both infrastructure and disposable inventory—living components that you can shut down, revive or replace as needed.

The idea has a long history in space planning. In the 1960s and 1970s engineers sketched mechanical and botanical systems for closed environments; more recently, synthetic biology reframed those sketches as biochemical engineering problems. NASA's more recent work on microbial biomanufacturing and the BioNutrients concept crystallized a practical argument: engineered microbes can be simple, shelf-stable payloads aboard a mission and then, when supplied with in-situ resources like CO_2, water and regolith-derived minerals, produce a continuous stream of useful products. That shift—from bringing food and drugs to making them on Mars—changes how mission planners weigh mass, risk and autonomy.

Concrete demonstrations on orbit have made the hypothesis tangible. Small, ruggedized DNA sequencers and compact PCR machines reached the International Space Station years ago, enabling real-time monitoring of microbial communities and the health of engineered strains aboard an actual habitat. BioSentinel, a compact deep-space biology experiment, illustrated another vital point: organisms can be prepared in long-lived stasis and revived by a timed rehydration to report on environmental hazards such as radiation. Those technical building blocks—compact diagnostics, stasis preservation, automated readouts—are the scaffolding that allows living systems to be treated as mission assets rather than uncontrolled biology.

Designing such systems requires thinking in three linked registers: the biochemical capabilities of organisms, the physical environment in which they must operate, and the control logic that keeps the loop steady. Each register offers opportunities and surprises.

On the biochemical side, modern tools let engineers redesign metabolism with extraordinary focus. A yeast strain can be retooled to overproduce a particular vitamin; a photosynthetic microbe can be tuned to secrete amino acids rather than stockpiling biomass. Those changes can be modest genetic edits or wholesale pathway rewires guided by machine-learning models and automated evolution. But on Earth the lab bench shares a predictable set of boundary conditions: gravity, convection, abundant water and a familiar radiation background. Space is different.

Microgravity alters fluid behavior: without buoyancy-driven convection, diffusion and capillary forces dominate. This matters for life support because transport controls encounter rates—between microbes and nutrients, between phage—bacteria hosts, between metabolites and enzymes. Experiments flown in low Earth orbit have shown that physical constraints reshape ecological interactions. For instance, phage—bacteria coevolution flown aboard the ISS produced different evolutionary paths than matched ground controls: infection dynamics were delayed, and both virus and host accumulated mutations that reflected selection in the altered transport milieu.

Such findings matter for two reasons. First, they remind designers that strains validated on Earth can behave differently in orbit or on Mars; second, they suggest that space itself can be an accelerator of evolution, yielding variants with unexpected properties—some useful, some risky.

Radiation adds a second twist. Outside Earth's protective magnetosphere, higher fluxes of ionizing particles create distinct mutation signatures and can damage biomolecules in ways not typically encountered in terrestrial labs. The BioSentinel mission used desiccated yeast with a DNA repair-deficient strain as a biological dosimeter, demonstrating how living assays can quantify radiation effects in situ. For life-support organisms, background mutation rates and the nature of damage shape long-term stability. Designing for robustness means engineering repair pathways, redundancy and, when appropriate, controlled dormancy: keep the working population small and revivable rather than large and constantly metabolically active.

The third register—control logic—is the leather in the stitching between hardware and wetware. Sensors, microfluidics and automated controllers translate analytic signals (optical density, dissolved gases, metabolite concentrations) into concrete actions (feed this reactor glucose; switch this strain to a starvation program; route off-gas to scrubbers). Modern approaches borrow from industrial control, but biology forces different trade-offs. A sudden reactor upset can cascade because organisms grow exponentially; conversely, a well-tuned system can recover from perturbations if the controllers act early. Increasingly, the control layer resembles a black box of models and policies: machine-learning models trained on ground and space data predict reactor outcomes; closed-loop evolution platforms can propose new strain variants that are synthesized and tested automatically. For missions where human attention is a scarce resource, automation is not a convenience—it is a necessity.

All of this sounds promising until one remembers the twin constraints of planetary protection and human safety. Sending living organisms to another planet brings ethical, legal and scientific obligations. Planetary protection protocols, intended to preserve scientifically pristine targets and prevent forward contamination, limit how and where biological systems can be used. Engineers counter with two strategies: confine biology to controlled habitats, and design organisms with genetic firewalls. Containment can be mechanical—airtight bioreactors, multiple redundant barriers, strict interface protocols—or biological. Biological containment uses dependency on synthetic nutrients, engineered auxotrophy, programmed kill-switches or dependency on non-standard genetic chemistries that cannot be replicated in the wild. Xenobiological systems—organisms that require synthetic base pairs or alternative backbone chemistries—offer attractive theoretical safety guarantees: escape and reproduction in a natural environment would be implausible because the host environment lacks the synthetic building blocks the organism needs. Those measures are not perfect, but they shift the risk calculus.

A useful way to see closed-loop design is as a circular economy, where resources flow in a loop rather than out to space. Near the center are reactors and processors; around the circle are inputs and outputs. To make this concrete, picture a Mars habitat where the main inputs include human waste streams (urine, greywater, food residues), atmospheric CO_2, sunlight and regolith-derived minerals. Cyanobacterial modules capture sunlight and CO_2 to produce oxygen and fixed carbon. Heterotrophic consortia ferment waste into short-chain organic acids, which are then fed to production strains: engineered bacteria or yeast that synthesize protein-rich biomass, vitamins or pharmaceuticals on demand. Microbial electrosynthesis units use electrons from renewable power to reduce CO_2 into more complex molecules—feedstocks for bioplastic production or rocket propellant precursors. Solid by-products are used to amend regolith for plant cultivation, creating a gradual build-up of soil-like material for expanded agriculture. Sensors continuously monitor the loop: oxygen partial pressures, nitrate levels, pathogen signatures. When a sensor flags an anomaly, the control system reallocates resources, isolates the affected module and, if necessary, launches a recovery protocol that may include dilution, heat treatment or selective antibiotics.

A diagram makes the idea less abstract; the circular flow clarifies the choreography of materials and functions. The visual aft of this description shows distinct arrows for waste conversion, photosynthetic capture, chemical-electrical interfaces and material production—each arrow labeled with the key transformation and the reactor class that performs it.

Real missions already test pieces of this vision. Early stations proved that routine microbial monitoring and sequencing are feasible in orbit; the technology to read DNA aboard a habitat has been demonstrated. BioNutrients-style concepts explored the possibility of shelf-stable precursor feeds and engineered microbes that, when activated, produce antioxidants or micronutrients tailored to a crew's dietary requirements. Modular microfluidic bioreactors, developed for fast biological assays, can also serve as compact production units: a set of cartridges containing lyophilized cells that revive when hydrated and run through a program of growth, induction and harvest.

Yet making these systems reliable requires wrestling down a set of sobering engineering realities. First, biology displays both resilience and brittleness: a population can absorb a range of perturbations, but certain stresses—contamination, unanticipated mutations or a malfunctioning sensor that misroutes feeds—can produce abrupt failures. Designing for graceful degradation is therefore essential: systems should continue to provide basic life support (oxygen scrubbing, water purification) even if production of luxury items (fresh food variety, complex pharmaceuticals) fails. Second, redundancy often trumps sophistication in remote settings; a simple parallel strain with partial capabilities may be more valuable than a single highly optimized but brittle organism. Third, metrics matter. On Earth, growth rate, yield and purity are common yardsticks; in space, stability, stall-revival behavior and minimal maintenance overhead become equally important. A strain that gives high yield in the lab but requires frequent manual correction is a poor match for an automated habitat.

Beyond engineering, there are human judgments to consider. Which products should be made in situ versus brought from Earth? How much biological autonomy is acceptable in a crewed habitat? Who decides containment thresholds? These questions are technical, political and moral. They intersect at the point where a habitat's biological systems meet the environment outside; even a perfectly contained loop can be a reputational risk if an escape event suggests possible contamination of a celestial body.

Those concerns fuel two research directions that are especially productive. One is the development of minimal, purpose-built production strains with narrow functional envelopes: organisms engineered to do one task and nothing else. Narrowness reduces unexpected behavior and simplifies monitoring. The other is a focus on reversible dormancy and rapid revival. A desiccated production cartridge that sits inert for months and then reliably revives when hydrated is easier to certify for planetary protection than a continuously active biomass. BioSentinel's approach—stasis enhanced by desiccation and timed reactivation—shows how biological payloads can be made temporally predictable.

An operational vignette helps make these trade-offs concrete. Consider a crewed mission en route to Mars that plans to produce vitamins and a particular antibiotic on demand. Pre-launch, engineers load cartridges of lyophilized yeast engineered to overproduce provitamin compounds and a separate bacterial strain designed to synthesize the antibiotic precursor. During transit, the production cartridges remain desiccated; monitoring systems watch for drift in habitat conditions. Midway through the journey, radiation events cause increased DNA damage in a small fraction of the on-board microbiome. Sensors detect altered metabolite profiles from a test cartridge and the control system automatically quarantines affected modules and switches to backup cartridges. Crew medical officers, informed by on-board sequencing data, decide whether to authorize use of antibiotics produced by the revived cartridge. The modular design, redundancy and automated diagnostics allowed the mission to isolate the problem and keep primary life-support functions intact while recovering specialized production capacity.

This interplay of risk management, biological design and engineering pragmatism is why the phrase *interplanetary Petri dish* is both apt and misleading. Apt because enclosed habitats are literally microcosms in which microbial populations grow, interact and evolve; misleading because the goal is not to cultivate uncontrolled ecology but to harness and constrain biological function with the precision once reserved for mechanical systems. The long-term promise is compelling: habitats that reduce resupply dependence, increase crew health autonomy and enable expansive exploration without the prohibitive mass penalty of carrying everything from Earth. The path there will be iterative—small, instrumented steps that prove components, expose surprises and refine the interplay between wetware and hardware.

In the meantime, practical steps tighten the gap between concept and reality. Standardized modular reactors, cartridge-based strain exchange, robust stasis protocols, onboard sequencing and analytics, and biological containment strategies all reduce operational risk.

International norms and planetary protection guidance shape where and how living systems can be deployed. And above all, experiments in orbit and on analog Mars sites on Earth will continue to reveal the unexpected ways that physics and evolution conspire to surprise careful plans. The lesson for designers is familiar to anyone who has worked with living systems: build in humility, automate where you can, and accept that a little redundancy buys a great deal of safety.

8.2 Evolution at the Speed of Silicon

It once took years to coax a single enzyme to do something new. Researchers would mutagenize a gene, grow colonies on plates, test a handful of candidates, and then iterate. The rhythms of that work—pipetting, waiting, discarding—shaped the tempo of biological discovery. That tempo is changing. Computers now act as both microscopes and test-tubes: they let researchers send millions of candidate designs through a rapid, inexpensive virtual sieve before a single pipette tip touches DNA. The result is a different kind of laboratory, one where evolution can happen thousands of times inside a cluster of servers in the time it takes a conventional experiment to finish a single day's growth.

A digital twin in biology is a software mirror of a living system: a mechanistic model of metabolism, a probabilistic emulator of folding behavior, or a machine-learning predictor of sequence-to-function relationships. Digital twins vary in fidelity. Some capture basic thermodynamics or enzyme kinetics; others are black-box statistical models trained on huge datasets of sequences and measured activities. The central promise is simple: evaluate more designs more quickly, prune the hopeless options, and focus expensive wet-lab resources on the most promising leads. What used to be a slow hunt through a dense forest of possibilities becomes, ideally, a guided trek along well-lit paths.

One of the clearest proofs of concept for computational evolution did not begin with proteins but with programs. The digital evolution platform Avida allowed self-replicating computer programs to compete, mutate and adapt inside a sandbox. Researchers such as Charles Ofria and Richard Lenski showed that these digital organisms could evolve complex logic functions through a series of small, selectable steps. That demonstration mattered because it separated a common intuition—complex features require complex, improbable leaps—from a testable claim. Under selection and variation, complexity can emerge incrementally. The lesson translated to biology: if you can define a fitness metric and provide variation, selection will often find paths that human designers did not predict.

On the wet-lab side, directed evolution—iterative rounds of mutagenesis and selection—has long been a practical route to new biomolecules. Frances Arnold won a Nobel Prize in 2018 for pioneering this method, which treats evolution as an engineer's tool: generate diversity, select for improved function, repeat. Directed evolution has produced enzymes that work in industrial conditions, catalysts for clean chemistry and components for novel biosensors. But the traditional loop is labor- and time-intensive. That is where computation promises leverage. Machine-learning-guided directed evolution (MLDE) replaces much of the blind search with informed proposals. Instead of sampling randomly or exhaustively, an algorithm learns from early measurements which sequence features correlate with success and then suggests new variants that are more likely to perform well.

A typical MLDE pipeline looks like a conversation between bench and server. Researchers build a modest library of variants and measure their activities. Those measurements train statistical models—sometimes simple regression, sometimes deep neural networks—

that estimate function from sequence. The model proposes a set of new candidates expected to improve the metric. The lab builds and tests them, and the results feed the next round. This closed loop compresses the number of wet-lab cycles required to reach high performance. In several documented cases, ML-guided searches achieved in a few rounds what formerly took dozens.

Generative models add another capability: rather than proposing tweaks to an existing protein, they imagine new structures from scratch. By learning patterns across thousands of protein sequences and structures, generative tools can create novel folds or scaffolds that satisfy design constraints—binding a ligand, catalyzing a reaction, or remaining stable under unusual conditions. Models such as those inspired by advances in protein structure prediction turn inversion problems into tractable design exercises: specify the function or geometry, and the model proposes sequences likely to realize it. These proposals are not guaranteed to work, but they broaden the design space in ways that simple mutation-based searches cannot.

That breadth is valuable because biological fitness landscapes are treacherous. Imagine a three-dimensional map where height equals performance and the horizontal coordinates are sequence choices. Local peaks—good solutions—abound, but the highest mountain may be separated from a hilltop by a deep valley of non-functional intermediates. Simple greedy search climbs the nearest slope and may get stuck. Computational search strategies can take larger, more informed leaps, identify promising ridgelines and navigate around valleys by proposing coordinated changes that preserve function while enabling distant improvements. Yet the maps themselves are only approximations. Epistasis—where the effect of one mutation depends on the presence of others—makes many landscapes rugged and context-dependent. A model trained on one region may mislead when asked to extrapolate.

The gap between virtual promise and biological reality is not merely theoretical. There are striking success stories and instructive failures. On the success side, teams have used MLDE to dramatically reduce the experimental volume needed to improve enzyme thermostability or substrate specificity. On the other side, models have produced candidates that look excellent on paper but fail in the bench because they ignored unmodeled constraints: post-translational modifications, intracellular degradation, expression bottlenecks, or interactions with chaperones. The lesson is practical: models are powerful but incomplete; their predictions must be validated by experiments that measure the right phenotypes under realistic conditions.

Bridging that gap has given rise to hybrid platforms—biofoundries—where automation links design, build and test in tight cycles. Robots assemble DNA, set up microplate assays, and return quantitative measurements to the design software. In such an integrated loop, a machine can propose thousands of variants, physically construct hundreds, and test dozens in a single day. Coupling high-throughput measurement with rapid modeling accelerates learning: the system discovers which features matter and refines its next proposals. When these platforms are paired with generative models or digital twins that incorporate mechanistic knowledge, the results can leapfrog what was possible with manual cycles.

There is also an intriguing middle ground: deliberately engineered genomes that include mechanisms for rapid, controlled diversification. The Synthetic Yeast Project (Sc2.0) introduced the SCRaMbLE system—an inbuilt recombination tool that can scramble segments of a synthetic chromosome to generate genotypic diversity on demand. Instead of designing each variant by hand, researchers can release SCRaMbLE in a controlled context, select for desirable traits, and then sequence winners to learn which rearrangements worked. SCRaMbLE is a wet-lab analogue of stochastic search and pairs naturally with computational analysis: sequence winners, learn patterns, refine models, and iterate.

Beyond engineering speed, computation changes how scientists think about contingency and reproducibility. Traditional wet-lab evolution experiments could generate interesting but often idiosyncratic results: one lab's replicate might find a particular mutation that another's did not. Digital evolution and high-throughput automation make it possible to run thousands of replicates with controlled variation in parameters. That scale lets researchers quantify the role of chance versus necessity. Is a particular adaptation a single rare event, or one of many likely outcomes given a class of selection pressures? Answering that question matters both for basic science and for design: predictable paths are easier to harness; contingent paths demand different risk management.

Speed and scale also affect ethical and safety considerations. Lowering the bar to design has a democratizing effect: more teams, including small labs and startups, can pursue ambitious biological engineering projects. That diversification fosters innovation, but it raises governance questions. Who audits the models? How are training datasets curated to avoid embedding dangerous shortcuts? A model that suggests sequences likely to increase a virus's infectivity would be as much a product of its data and objective as of clever algorithms. Responsible practice requires not only technical controls—model explainability, red-team testing, access restrictions—but also community norms and oversight that recognize where computational power transforms biological risks.

The limits of prediction come into sharp relief in environments far from those in which models were trained. A protein optimized for performance at room temperature may denature in the cold of space. A metabolic pathway tuned for rich laboratory media may collapse in nutrient-poor conditions. For applications such as off-world life support, designers must account for novel physical constraints—radiation, microgravity, low-pressure atmospheres—and for interactions with hardware. That necessity encourages two converging strategies. One is robustness by design: prefer solutions that perform adequately across a broad range of conditions rather than exquisitely in a narrow niche. The other is the incorporation of environmental physics into the digital twin; models that simulate transport, radiation-induced damage, or resource scarcity produce predictions that are more relevant to real-world deployment.

There is a cultural shift too. Where once molecular biology prized intuition and artisanal skill, a new practice values data curation, model validation and software engineering. Bench scientists learn to speak the language of loss functions and cross-validation; computational experts learn the messiness of clonings, such as off-target effects and batch-to-batch variability. The most productive teams are often those that sit at the interface—people who can recognize a model's artefact in an experimental readout and an experimentalist who understands why a suggested sequence might cause expression problems.

Historical anecdotes illuminate how these changes unfold. Consider the early days of the Avida experiments: a small group of researchers watched, sometimes incredulously, as programs without biochemistry evolved logic functions that their human designers had assumed required complex engineering. That surprise carried a message about what selection can discover. Decades later, directed evolution practitioners watched analogous surprises in the wet lab—enzymes finding catalytic tricks that neither computational chemists nor intuition had foreseen. Today, when a generative model proposes a protein fold no human has invented, the reaction echoes those earlier astonishments. The common thread is humility: nature, digital or biochemical, often finds clever, unexpected solutions if given variation and selection.

Practical outcomes of this accelerated paradigm are already visible.

Enzyme catalysts that once required months to improve can be optimized in weeks; therapeutic antibody candidates can be narrowed from thousands to dozens before animal testing; metabolic pathways can be redesigned to yield higher titers with fewer metabolic side effects. In each case, computation did not remove the need for careful experiments—it multiplied the experimental leverage. The laboratory became a selective filter rather than a random sweep.

As the machinery of design grows smarter, so do the ethical stakes. Predictive models can be repurposed, intentionally or accidentally, to create hazardous variants. Preventing misuse requires layered solutions: technical safeguards in software and hardware, norms around data sharing and model publication, and thoughtful regulation that balances innovation with safety. In practice that means public and private actors working together to define red lines, implement practical controls and provide channels for responsible research.

Finally, the partnership between silicon and life reframes what it means to do biological engineering. Evolution used to be an unpredictable story told slowly; now it can be rehearsed and interrogated inside a computer. That rehearsal does not make outcomes fully predictable—chance and context continue to matter—but it changes what is possible. Engineers who once feared getting lost in the space of sequences can now navigate with maps sketched by computation and refined by experiment. The maps are imperfect, but they are improving. With them comes new power: to design organisms that help sustain life in harsh environments, to discover enzymes that revolutionize industry, and to probe the deepest questions about how function emerges from the code of life. Alongside that power sits responsibility: to use predictive tools wisely, to validate designs thoroughly, and to maintain humility about what remains unpredictable.

8.3 Beyond the ACGT Cipher

Life as we know it writes itself with four letters: A, C, G and T. The simplicity of that alphabet—paired bases forming the familiar double helix—masked a profound possibility: that the chemical underpinnings of heredity might be a historical accident rather than an absolute constraint. If DNA is a successful solution rather than the only solution, then it becomes conceivable to redesign the alphabet itself. That redesign is the province of xenobiology, a field that treats the chemical grammar of life as malleable. Changing the letters changes the story you can tell with genes: new amino acids, novel backbones, and perhaps most importantly for safety, a code that does not speak the language of natural organisms.

The first practical motivations for tinkering with genetic chemistry were pragmatic. Natural biology is superb at many things-replication, catalysis, self-assembly—but it does not produce every useful molecule on demand, nor does it always behave politely outside its evolved contexts. Synthetic biologists wanted tools that could expand the palette of chemistry accessible to living systems and, at the same time, create forms of containment: organisms that could flourish inside laboratory hardware but not in the wild. These two goals-capability and containment-often point to the same solution: make life that depends on building blocks absent from nature.

One of the most visible early successes came from the work on alternative genetic polymers, collectively called XNAs (xeno-nucleic acids). In the 2010s, researchers demonstrated that information could be stored and propagated in polymer backbones different from DNA and RNA. Those experiments did more than show a neat chemical trick; they undermined an assumption long held outside the chemistry community: heredity and evolution do not uniquely require the phosphate-sugar backbone of DNA. With the right polymerases-enzymes adapted or engineered to copy between chemistries-XNAs could be synthesized, copied and even selected for functional folding. The implication is striking: the property of being evolvable does not demand Earth's specific molecular choices.

Broadening the alphabet took two parallel routes. One approach added new base pairs to the familiar backbone; the other replaced the backbone altogether. Adding synthetic base pairs-un-natural nucleotides that pair with one another but not with the canonical bases-creates an extended code that sits awkwardly, and usefully, beside natural DNA. A notable achievement along this path was the creation of semi-synthetic organisms that stably maintained an extra base pair inside their genome. The pair known as dNaM–dTPT3 became a practical demonstration that living cells can store and retrieve information beyond the four-letter alphabet, transcribe it into RNA, and use it during translation to incorporate non-canonical amino acids into proteins. That work showed two things at once: non-standard nucleotides can be made to work inside a cell, and those novel letters open the door to proteins that carry chemical functionality nature did not invent.

A separate but conceptually related thread produced a striking laboratory demonstration of an eight-letter genetic system, nicknamed *hachimoji* DNA. Chemists designed four orthogonal base pairs that could form a stable double helix with predictable thermodynamics, and then showed that this expanded helix could be transcribed into RNA and fold into functional structures, such as aptamers. The achievement was both pragmatic and philosophical. Pragmatically, it demonstrated that the information density of genetic polymers can be increased and that new base pairs can be designed to play nicely with the structural constraints of double helices. Philosophically, it suggested that the DNA alphabet is not sacred; it is one design among many that can support molecular information storage and function.

Why does any of this matter beyond intellectual curiosity? For engineers it matters because an expanded genetic system is a new design space. Each extra letter multiplies the possible codons and, potentially, the number of distinct amino acids that a ribosome could assemble into a protein. That expansion creates opportunities: enzymes whose active sites use non-standard residues to stabilize transition states, materials with covalent linkages unavailable to standard amino acids, or therapeutics that resist natural degradation pathways. In short, xenobiology promises both novel function and increased design control.

Safety considerations have been a central motivation for many xenobiological efforts. A genetic firewall arises when an engineered organism requires building blocks absent from natural ecosystems-synthetic nucleotides, exotic amino acids, or a non-natural backbone-to replicate and express essential genes. If such an organism escapes a controlled environment, the absence of its required substrates should prevent it from surviving or exchanging functional genetic material with native organisms. This strategy resembles a digital firewall: set an authentication token that only your authorized systems possess. In biology, the token is a chemical that the outside world does not normally provide.

That logic underpinned early proposals to use dependency on synthetic nucleotides as an effective containment strategy for organisms sent beyond the laboratory—whether to a field site on Earth or to a habitat on another world. For environments such as a sealed Mars habitat, where human planners supply everything inside the walls, requiring a synthetic nucleotide makes practical sense: the habitat controls supply chains, and the risk of an escaped organism establishing itself on Martian soil is reduced because its biochemical lifeline would be absent outside the habitat.

The promise of a chemical firewall is powerful but not absolute. Engineering dependency is not the same as guaranteeing impossibility. Evolution is an opportunist. Given enough mutational time and selection pressure, organisms can find workarounds: mutate polymerases to accept natural nucleotides, recruit host enzymes, or scavenge precursors from unexpected sources. Horizontal gene transfer could, in theory, move functional modules between organisms in ways that bypass containment assumptions. Worse, the very act of engineering orthogonal systems often requires creating specialized enzymes—polymerases, transcription factors and ribosomes—that themselves are proteins encoded in DNA. Those proteins can, in principle, be modified or transferred. Containment by chemical dependency shifts the risk rather than eliminating it.

Historical experiments exemplify both the power and the fragility of these strategies. Creating a semi-synthetic organism required not only synthesizing nucleotides that paired reliably but also engineering nucleotide transporters and polymerases to import and replicate them inside a living cell. The early successes depended on careful supply control: the synthetic nucleotides were provided in culture, and removal arrested propagation. That controllability was reassuring. Yet follow-up work showed that without careful oversight, selective pressures could favor variants that reduced their dependency or found alternative metabolic routes. Each success thus spawned new technical questions about long-term stability.

Parallel work on XNAs emphasized a different set of issues. Because XNAs use alternative backbones, they are intrinsically less likely to interact compatibly with natural enzymes. That chemical mismatch can be a powerful barrier to horizontal information flow: DNA polymerases typically won't copy XNA, and natural nucleases won't readily degrade it. But engineering the polymerases that do copy XNA demands deliberate intervention—creating enzymes that bridge the worlds. Those engineered polymerases are the very tools that make

XNA useful, and they are also potential weak points in a containment philosophy if their function escapes the designed context.

The technical hurdles of xenobiology are substantial. Polymerases are finicky; their active sites evolved to handle the chemistry of natural nucleotides. Re-engineering them to accept non-standard sugars, altered backbones or synthetic bases often requires rounds of directed evolution—a process that itself must be carefully contained. Similarly, ribosomes and tRNA synthetases must be adapted or duplicated to translate expanded codons into non-canonical amino acids. Orthogonal translation systems-parallel protein synthesis pathways that operate alongside the cell's native machinery—have been built, but they complicate cellular economy: energy use, error rates and interaction networks all shift, sometimes in unpredictable ways.

Those unpredictable effects are not merely technical annoyances; they shape the risk profile of xenobiological systems. An orthogonal pathway that competes for shared metabolites can reduce cellular fitness and thereby select for mutations that disable the orthogonal components. In a closed system where the orthogonal function is valuable-production of a drug, for instance—selection pressures may favor robustness of the orthogonal machinery. Robustness, in turn, might make the system less dependent on human-controlled inputs. That trajectory illustrates a tension at the heart of design-for-containment: making an organism robust enough to be practically useful often edges it away from perfect containment.

Beyond engineering and biosafety, xenobiology raises philosophical and ethical questions. If we create forms of life whose chemistry is meaningfully different from Earth life, what moral status, if any, attaches to those organisms? Do they count as life in the same way as bacteria and humans? Practically, the issues center more on responsibility than on metaphysics. Novel chemistries complicate environmental risk assessment and legal frameworks that regulate release, transport and intellectual property. They demand updates to planetary protection rules and public dialogue about acceptable experiments in field settings.

There are also deep scientific payoffs that justify the careful, measured pursuit of xenobiology. Exploring alternative chemistries probes hypotheses about the contingency of life: were DNA and RNA inevitable outcomes of chemistry, or simply one solution among many? Demonstrating functional heredity and evolution in XNA systems supports the idea that life's core properties-information storage, replication and selection-are abstract and potentially realizable in multiple chemistries. That insight bears directly on astrobiology. If heredity can be supported by many chemical scaffolds, then the probability space for life in the universe widens.

Concrete stories anchor these abstract debates. When scientists working on *hachimoji* DNA reported that an eight-letter system could form predictable helices and be transcribed into functional RNAs, the community reacted with a mix of excitement and caution. Excitement because the experiment opened new design possibilities; caution because the work showed how easy it is to imagine life-like chemistries that would defy current biosafety paradigms. Similarly, the early creation of a semi-synthetic organism carrying extra base pairs felt like a watershed: it was proof that the tape of life could be edited at a fundamental level. Both achievements also sparked sober conversations about governance. If one can code information in alien alphabets, how should society regulate organisms that use them?

Regulatory frameworks have begun to respond. Practical measures are technical (mandatory dependency on supplied synthetic building blocks, fail-safe kill switches, physical containment) and institutional (review boards, staged field trials, transparency in methods). Scientific practice too has changed: dual-use risk assessments are increasingly standard in grant proposals; journals and funding agencies scrutinize proposed work that could enable misuse. Those steps matter, but they are not foolproof. The pace of computational design-machine learning models that can propose efficient polymerases or plausible orthogonal enzymes-means that capability can outstrip oversight unless governance evolves in parallel.

A final, often-overlooked point is that xenobiology does not exist in isolation. Its tools intersect with other advances: automation accelerates the rounds of polymerase engineering needed to make XNA practical; computational protein design proposes aminoacyl-tRNA synthetases that can attach new amino acids; minimal-genome efforts reveal how much of life's wiring remains poorly understood even in pared-down organisms. These intersections are a strength: they create synergies that make ambitious ideas feasible. They are also a source of compounded uncertainty: more moving parts mean more pathways to unexpected outcomes.

The promise of xenobiology is therefore a study in contrasts. It is a technology that can simultaneously expand the chemical reach of biotechnology and erect new layers of safety by altering the basic grammar of heredity. It is also a technology that creates fresh avenues for failure if human systems—technical, institutional and ethical—do not keep pace. The task for practitioners and society is not to choose between innovation and caution but to design experiments, norms and institutions that allow exploration while minimizing the probability of harm. Doing that well requires technical rigor, transparent deliberation and a willingness to accept limits on certain experiments until robust safeguards exist.

www.ingramcontent.com/pod-product-compliance
Lightning Source LLC
LaVergne TN
LVHW022211200926
848481LV00006B/1926